SWARM INTELLIGENCE
Principles, Advances, and Applications

SWARM INTELLIGENCE
Principles, Advances, and Applications

Aboul Ella Hassanien
Eid Emary

CRC Press
Taylor & Francis Group
Boca Raton London New York

CRC Press is an imprint of the
Taylor & Francis Group, an **informa** business

CRC Press
Taylor & Francis Group
6000 Broken Sound Parkway NW, Suite 300
Boca Raton, FL 33487-2742

© 2016 by Taylor & Francis Group, LLC
CRC Press is an imprint of Taylor & Francis Group, an Informa business

No claim to original U.S. Government works

ISBN-13: 978-1-4987-4106-4 (hbk)
ISBN-13: 978-0-3677-3754-2 (pbk)

Visit the Taylor & Francis Web site at
http://www.taylorandfrancis.com

and the CRC Press Web site at
http://www.crcpress.com

Contents

List of Figures

List of Tables

Preface

Swarm intelligence and bio-inspired computation have become very popular in recent years. Optimization in general and swarm-based intelligence in particular are becoming a basic foundation for many modern applications in different disciplines. For its simplicity and flexibility swarm intelligence has found its way in engineering applications. This book presents a brief introduction to the mathematical optimization with special attention to swarm intelligence and its applications, divisions, variants, and hybridizations. Basic concepts related to swarm intelligence are presented such as randomness, random walks and chaos theory as a preface for many of the variants and enhancements in the literature for the basic swarm optimization methods.

This is a book for researchers and students who are interested in understanding and solving swarm optimization problems and also to developing an understanding of new swarm optimization techniques and knowing their advantages and limitations that lead to a better understanding of their impact on various applications. The main goal of this book is to give a comprehensive description of the newest modern swarm algorithms including

- Bat algorithm (BA)

- Artificial fish swarm algorithm (AFSA)

- Firefly algorithm (FFA)

- Cuckoo search (CS)

- Flower pollination algorithm (FPA)

- Artificial bee colony (ABC)

- Wolf search algorithm (WSA)

- Gray wolf optimization (GWO)

For the individual optimization methods, we tried to standardize the variants and hybridizations and even the algorithms. The variants focus more on binary, discrete, constrained, adaptive, and chaotic versions of the different optimizers. Applications of individual optimizers mentioned in the literature are also presented. When presenting the different applications the focus was toward variable selection, and fitness function design to fit the different applications. The final part of the book is dedicated to exploring similarities, differences, weaknesses, and strengths of individual methods covered in the book. This final part tries to standardize the operator concepts and searching manner of the different algorithms covered. In our opinion, this book can provide a good base for swarm intelligence beginners as well as experts and can provide a complete view of new directions, enhancements, and hybridizations of the different optimizations.

A chapter-by-chapter description of the book follows.

Chapter 1 is an introductory chapter that aims at setting the environment for the whole book. This chapter starts with basic definitions and concepts related to optimization and its sources of inspiration. It then describes randomness as a basic building block for stochastic optimization and common random distributions that are commonly exploited in optimization. Moreover, it describes the principle of random/pseudo number generation using common random distributions. We cover the topics of random distribution and pseudo-random generation as it is the base for modifying and enhancing common optimizers. Random walk is a common concept in optimization and hence it is covered as well. Because chaos theory and chaos variables are becoming common as independent optimization tools or as a method for adapting optimizers' parameters, a brief discussion is given about the chaos theory.

Chapter 2 **Bat Algorithm (BA)** is a new meta-heuristic optimization algorithm derived by simulating the echolocation system of bats. It is becoming a promising method because of its good performances in solving many problems. This chapter discusses the behavior of bats and its variants including the discrete bat algorithm, binary bat algorithm, chaotic bat algorithm, parallel bat algorithm, bat for constrained problems, bat with Lèvy distribution, chaotic bat with Lèvy distribution, self-adaptive bat algorithm and an adaptive bat algorithm. Moreover, this chapter reviews the most hybrid evolutionary bat algorithm with other optimization techniques including differential evolution (DE), particle swarm optimization, cuckoo search, simulated annealing, harmony search and artificial bee colony. This chapter ends with a discussion of a real-world application as a case study of bat algorithms. Finally, an extensive bibliography is also included.

Chapter 3 **Artificial Fish Swarm (AFS)** is one of the best and new swarm optimization methods. The basic idea of the artificial fish swarm is simulated fish behaviors such as swarming, preying, followed by a local search of a fish individual for reaching the global optimum; it is chaotic, binary and parallel search algorithm. This chapter reviews the basic concepts of the artificial fish swarm algorithm and discusses its variants and hybridization with other optimization techniques. In addition, we show how the artificial fish swarm algorithm is applied to solve real-life applications such as selection of optimal cluster heads (CHs) locations in wireless networks and in community detection in social networks.

Chapter (4) **Cuckoo Search (CS)** is one of the latest biologically inspired algorithms, proposed by Yang and Deb in 2009 and Suash Deb. It is based on the brood parasitism of some cuckoo species. In addition, this algorithm is enhanced by the so-called Lèvy flights rather than by simple isotropic random walks. This chapter provides basic information about the cuckoo search algorithm and its behavior. In addition, it discusses its variants including discrete, binary, chaotic and parallel versions. A hybridization of the cuckoo search (CS) with other approaches is discussed in this chapter with differential evolution, scatter search, colony optimization, powell search, bat algorithm, particle swarm optimization, Levenberg–Marquardt and quantum computing. Reviews of a real-world application as well as results of two cases in feature selection and a modified version of a cuckoo search for solving a convex economic dispatch problem. Finally, an extensive bibliography is also included.

Chapter (5) **Firefly Algorithm (FA)** is a nature-inspired stochastic global optimization method that was developed by Yang [1]. This algorithm imitates the mechanism of firefly mating and the exchange of information using light flashes. This chapter presents the main concepts and behavior of fireflies, the artificial FFA, and the variants added to the basic algorithm including the discrete firefly, binary firefly, chaotic firefly, parallel firefly, firefly for constrained problems, Lèvy flight firefly, intelligent firefly, Gaussian firefly, network-structured firefly algorithm and firefly with adaptive parameters. Moreover, this chapter reviews the most hybrid evolutionary firefly algorithm with other optimization techniques including harmony search, pattern search, learning automata firefly algorithm and ant colony optimization. This chapter ends by discussing a real-world application as a case study of the firefly algorithm. Finally, an extensive bibliography is also included.

Chapter (6) **Flower Pollination Algorithm** was developed by Xin-She Yang in 2012, inspired by the flower pollination process of flowering plants. The flower pollination algorithm has been extended to multi-objective optimization with promising results. This chapter provides an introduction to the flower pollination algorithm and its basic characteristics. In addition, flower pollination algorithm variants are presented and its hybridization with other optimization techniques. The chapter reviews some real-life applications with the flower pollination algorithm applied. Moreover, feature selection that is critical for a successful predictive modeling exercise is introduced based on flower pollination algorithm.

Chapter (7) **Artificial Bee Colony Optimization** was introduced by Karabora in 2005. In this chapter, an overview of the basic concepts of artificial bee colony optimization, its variants, and hybridization with other optimization techniques is presented. In addition, different application areas are discussed, where the artificial bee colony optimization algorithm has been applied. In addition, reviews of the results of artificial bee colony on retinal vessel segmentation and the memetic artificial bee colony for integer programming are discussed.

Chapter (8) **Wolf-Based Search Algorithms** is a new bio-inspired heuristic optimization algorithm that imitates the way wolves search for food and survive by avoiding their enemies. It was proposed by Rui Tang in 2012. This chapter reviews the wolf search optimizers algorithm and its variants as well as reviews some real-life applications. In addition, it shows how wolf-based search algorithms are applied in a feature selection problem.

Chapter (9), presents a high-level discussion of the presented algorithms in this book and differentiates among them based on different criteria. It identifies similarities and differences in criteria setting. These criteria are according to the swarm guide, the probability distribution, the number of behaviors, exploitation of positional distribution of agents, number of control parameters, generation of completely new agents per iteration, exploitation of velocity concept in the optimization, and the type of exploration/exploitation used. This discussion may help in finding weak and strong points of each algorithm and may help in presenting new hybridizations and modifications to further enhance their performance or at least help us choose among this set to handle specific optimization tasks.

Additional material is available from the CRC website: http://www.crcpress.com/product/isbn/9781498791064

1 Introduction

Optimization is everywhere, from engineering applications to computer science and from finance to decision making. Optimization is an important paradigm itself with a wide range of applications. In almost all applications in engineering and industry, we are always trying to optimize something, whether to minimize the cost and energy consumption, or to maximize the profit, output, performance, and efficiency [1].

Mathematical optimization is the study of such planning and design problems using mathematical tools [2]. It is possible to write most optimization problems in the generic form as follows:

$$\min_{x \in \Re^n} f_i(x), \ (i = 1, 2, ...M) \tag{1.1}$$

$$Subject\ to\ h_j(x) = 0, \ (j = 1, 2, ...J) \tag{1.2}$$

$$g_k(x) \leq 0, (k = 1, 2, ..., K) \tag{1.3}$$

Where M is the number of cost or objective functions to be optimized $f_i(x)$ is the objective function representing objective i and x is the design variables to be guessed by the optimization algorithm and its size is n.

$\Re^n$ is the space spanned by the decision variables and is called the design space or search space. The equalities for h_j and inequalities for g_k are called constraints.

The optimization problem with $M = 1$ is called single objective optimization, while optimization at $M > 1$ is called multi-objective optimization. If no constraints are imposed on the problem $J = 0$ and $K = 0$, the optimization is said to be an unconstrained problem, while at $J > 0$ or $K > 0$, the problem is said to be constrained problem.

The optimization problem can be classified according to the linearity/nonlinearity of objective and/or constrains. When all f_i, h_j, g_k are nonlinear, the optimization problem is said to be nonlinear optimization problem. When only the constrains h_j and g_k are nonlinear, the optimization problem is said to be a nonlinearly constrained problem. When both h_j and g_k are linear, the problem is said to be a linearly constrained problem. It is worth mentioning that in some optimization tasks, some of the variables are linearly related to the objective function and others are nonlinearly related to it.

Classification of an optimization algorithm can be carried out in many ways. A simple way is to look at the nature of the algorithm, and this divides the algorithms into two categories: deterministic algorithms and stochastic algorithms [2]. Deterministic algorithms follow a rigorous procedure, and the path and values of both design variables and the functions are repeatable. Hill-climbing is an example of a deterministic algorithm, and for the same starting point, it will follow the same path.

Stochastic algorithms always have some randomness. Thus, intermediate and final solutions will differ from time to time. Particle swarm optimization (PSO) is an example of such non-repeatable algorithms. Based on the number of agents used to search the search domain, optimization algorithms can be classified as single agent and multiple agent optimizers [1]. Simulated annealing is a single-agent algorithm with a zigzag piecewise trajectory, whereas genetic algorithms, PSO, and firefly algorithms are population-based algorithms. These algorithms often have multiple agents, interacting in a nonlinear manner.

Depending on the sources of inspiration, optimization can be classified as bio-inspired, nature-inspired algorithms, and meta-heuristics in general [1]. Heuristic means by trial and error, and meta-heuristic can be considered a higher-level method by using certain selection mechanisms and information sharing [1]. Random search [2] is an example of such a category. Nature-inspired algorithms are algorithms drawing inspiration from nature such as artificial bee colony and genetic algorithms. Bio-inspired algorithms may be viewed as a branch of nature-inspired algorithms and it is inspire from a biological system such as a bee colony.

1.1 Sources of inspiration

Most new optimization algorithms are nature-inspired, because they have been developed by drawing inspiration from nature [5]. The highest level of inspiration sources are from *biology*, *physics*, or *chemistry*. The main source of inspiration is nature. Therefore, almost all the new algorithms can be referred to as *nature-inspired*. By far the majority of the nature-inspired algorithms are based on some successful characteristics of the biological system. Therefore, the largest fraction of the nature-inspired algorithms are biology-inspired, or bio-inspired [5].

Among the bio-inspired algorithms, a special class of algorithms has been developed based on swarm intelligence. Therefore, some of the bio-inspired algorithms can be called swarm intelligence-based. Examples of swarm-intelligence-based are cuckoo search [1], bat algorithm [1], and artificial bee colony [55]. Many algorithms have been developed by using inspiration from physical and chemical systems such as simulated annealing [14]. Some may even be based on music such as harmony search [15]. Fister et al. [5] divides all the existing intelligent algorithms into four major categories: *swarm intelligence(SI)-based*, *bio-inspired* (but not SI-based), *physics/chemistry-based*, and *others*.

1.1.1 Swarm intelligence–based algorithms

Swarm intelligence (SI) concerns the collective, emerging behavior of multiple, interacting agents that follow some simple rules [5]. Each agent may be considered as unintelligent, while the whole system of the multiple agents may show some self-organization behavior and thus can behave like some sort of collective intelligence. Many algorithms have been developed by drawing inspiration from the SI systems in nature. The main properties for the SI-based algorithms can be summarized as follows:

- Share information among the multiple agents.

- Agents have self-organization and co-evolution.

- It is highly efficient for its co-learning.

- It can be easily parallelized for practical and real-time problems.

1.1.2 Bio-inspired, but not swarm intelligence–based algorithms

The SI-based algorithms belong to a wider class of the algorithms, called the bio-inspired algorithms, as mentioned before. The SI-based algorithms are a subset of the bio-inspired algorithms, while the bio-inspired algorithms are a subset of the nature-inspired algorithms [5]. Thus, we can observe thatSI-based ⊂ bio-inspired ⊂ nature-inspired.

Many bio-inspired algorithms do not directly use the swarming behavior. It is better to call them bio-inspired, but not SI-based. For example, the genetic algorithms are bio-inspired, but not SI-based. The flower algorithm [1], or flower pollination algorithm, developed by Xin-She Yang in 2012, is a bio-inspired algorithm, but it is not an SI-based algorithm because the flower algorithm tries to mimic the pollination characteristics of flowering plants and the associated flower consistency of some pollinating insects.

1.1.3 Physics- and chemistry-based algorithms

For the algorithms that are not bio-inspired, most have been developed by mimicking certain physical and/or chemical laws, including electrical charges, gravity, river systems, etc. [5]. Examples of such algorithms, spiral optimization [12], water cycle algorithm [13], simulated annealing [14], and harmony search [15].

1.1.4 Other algorithms

Some developed algorithms are away from nature using various characteristics from different sources, such as social, emotional, etc. Examples of such a system are Imperialist competitive algorithm [16], league championship algorithm [17], social-emotional optimization [18].

1.2 Random variables

A random variable can be considered as an expression whose value is the realization or outcome of events associated with a random process [2]. A random variable is a function which maps events to real numbers. The domain of this mapping is called the sample space. Each random variable is represented by a probability density function to express its probability distribution. Random variables are mile stones for all stochastic optimization algorithm and hence we will discuss some of the random probability density functions.

1.2.1 Uniform distribution

The uniform distribution [20] is a very simple case with

$$f(x; a, b) = \frac{1}{b - a} \text{ for } a \leq x \leq b \tag{1.4}$$

where a, b is the distribution limiting range and is always set to [0 1]. The uniform distribution has expectation value $E(x) = (a + b)/2$ and variance $\frac{(b-a)^2}{12}$.

1.2.2 Normal distribution

Gaussian distribution or normal distribution is by far the most popular distribution, because many physical variables including light intensity, and errors/uncertainty in measurements, and many other processes obey the normal distribution.

The probability density function related to normal distribution is mentioned in equation 1.5.

$$p(x; \mu, \sigma) = \frac{1}{\sigma\sqrt{2\pi}} \exp[-\frac{(x - \mu)^2}{2\sigma^2}], -\infty < x < \infty \tag{1.5}$$

where μ is the mean and $\sigma > 0$ is the standard deviation. A Gaussian distribution with mean zero and standard deviation one, often known as a "standard normal" distribution, has the probability density function (PDF) shown in figure 1.1. In many connections it is sufficient to use this simpler form since μ and σ simply may be regarded as a shift and scale parameter, respectively.

1.2.3 Cauchy distribution

The Cauchy distribution is given by the following equation:

$$f(x) = \frac{1}{\pi} \cdot \frac{1}{1} + x^2 \tag{1.6}$$

x is defined in $-\infty < x < \infty$. It is a symmetric uni-modal distribution. Figure 1.2 is a sample figure of Cauchy distribution.

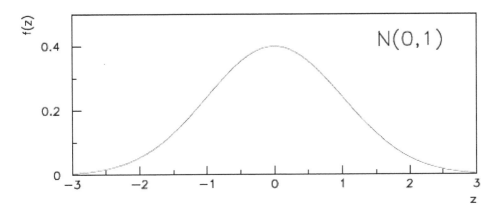

Figure 1.1: Normal distribution for N(0,1)

1.2.4 Poisson distribution

The Poisson distribution [20] is given by the following equation:

$$p(r; \mu) = \frac{\mu^r e^{-\mu}}{r!} \tag{1.7}$$

where r is an integer parameter ≥ 0 and μ is a real positive quantity. The Poisson distribution describes the probability to find exactly r events in a given length of time if the events occur independently at a constant rate μ. It is worth mentioning that when μ goes to ∞ the distribution tends to a normal one.

1.2.5 Lèvy distribution

Levy distribution [2] is a distribution of the sum of N identically and independently distribution random variables whose Fourier transform takes the following equation:

$$F_N(k) = \exp(-N|k|^\beta) \tag{1.8}$$

and the actual distribution from the inverse of Fourier is defined by the following equation:

$$L(s) = \frac{1}{\pi} \int_0^\infty \cos(\tau s) e^{-\alpha \tau^\beta} d\tau, (0 < \beta <= 2) \tag{1.9}$$

Where $L(s)$ is called Levy distribution with index β and α is always set to 1 for simplicity. Two special cases are $\beta = 1$, the above integral becomes the Cauchy distribution; where. When $\beta = 2$ it becomes the normal distribution. In this case, Levy flights become the standard Brownian motion. Equation in 1.9 can be expressed as an asymptotic

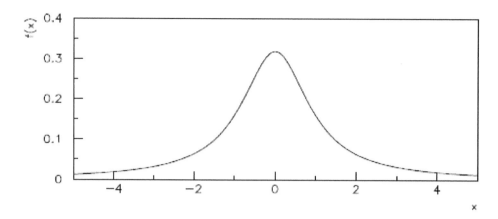

Figure 1.2: Cauchy distribution

series, and its leading-order approximation for the flight length results in a power-law distribution defined by the following form:

$$L(s) \; |s|^{-1-\beta} \tag{1.10}$$

which is heavy tailed with infinite variance for $0 < \beta < 2$. A simple version of the Levy distribution can be defined as

$$L(s, \gamma, \mu) = \begin{cases} \sqrt{\frac{\gamma}{2\pi}} \exp[-\frac{\gamma}{2(s-\mu)}] \frac{1}{(s-\mu)^{1.5}}, & \text{if } 0 < \mu < s < \infty \\ 0, & \text{otherwise} \end{cases} \tag{1.11}$$

where $\mu > 0$ is minimum step and γ is the scale parameter and as $s \to \infty$ we have the following form:

$$L(s, \gamma, \mu) \approx \sqrt{\frac{\gamma}{2\pi}} \frac{1}{s^{1}.5} \tag{1.12}$$

1.3 Pseudo-random number generation

Simulations requiring random numbers are critical in fields including communications, financial modeling, and optimization. A wide range of random number generators (RNGs) have been described in the literature [19]. They all utilize well-understood basic mathematical principles, usually involving transformations of uniform random numbers. Assuming suitably precise arithmetic, the RNGs can generally be configured to deliver random numbers of sufficient quality to meet the demands of a particular simulation environment.

RNGs aim to produce random numbers that, to the accuracy necessary for a given application, are statistically indistinguishable from samples of a random variable with an ideal distribution. RNGs can be classified as cumulative, accept-reject or composition techniques [20].

1.3.1 Cumulative technique

The most direct technique to obtain random numbers from a continuous PDF $f(x)$ with a limited range from x_{min} to x_{max} is to solve for x in the following equation:

$$\zeta = \frac{F(x) - F(x_{min})}{F(x_{max}) - F(x_{min})} \tag{1.13}$$

where ζ is uniformly distributed between zero and one and F(x) is the cumulative distribution (or as statisticians say the distribution function), for a properly normalized PDF. Thus

$$x = F^{-1}(\zeta) \tag{1.14}$$

The technique is sometimes also of use in the discrete case if the cumulative sum may be expressed in analytical form. Also, for general cases, discrete or continuous, e.g., from an arbitrary histogram, the cumulative method is convenient and often faster than more elaborate methods. In this case the task is to construct a cumulative vector and assign a random number according to the value of a uniform random number (interpolating within bins in the continuous case).

1.3.2 Accept/reject technique

A useful technique is the acceptance-rejection method where we choose f_{max} to be greater than or equal to $f(x)$ in the entire interval between x_{min} and x_{max} and proceed as follows:

1. Generate a pair of uniform pseudo-random numbers ζ_1, ζ_2.

2. Calculate x as $x = x_{min} + \zeta_1.(x_{max} - x_{min})$.

3. Determine y as $y = f_{max}.\zeta_2$.

4. If $y - f(x) > 0$ reject and go to 1. Otherwise accept x as a pseudo-random number from $f(x)$.

The efficiency of this method depends on the average value of $\frac{f(x)}{f_{max}}$ over the interval. If this value is close to one the method is efficient.

1.3.3 Composition technique

Let $f(x)$ be written in the form

$$f(x) = \int_{-\inf}^{\inf} g_z(x) dH(z) \tag{1.15}$$

where we know how to sample random numbers from the PDF, $g(x)$, and the distribution function $H(z)$. A random number from $f(x)$ is then obtained by

1. Generating a pair of uniform pseudo-random numbers ζ_1, ζ_2

2. Calculating z as $z = H^{-1}(\zeta_1)$

3. Determining x as $x = G^{-1}\zeta_2$, where G_z is the distribution function corresponding to the PDF $g_z(x)$.

1.4 Random walk

A random walk is a random process that consists of taking a series of consecutive random steps [2] as actually occurs in optimization. Let S_N denote the sum of each consecutive random step X_i, then S_N forms a random walk; see equation 1.16.

$$S_N = \sum_{i=1}^{N} X_i = X_1 + ... + X_N \qquad (1.16)$$

where X_i is a random step drawn from a random distribution. This relationship can also be written as a recursive formula as defind as follows:

$$S_N = S_{N-1} + X_N \qquad (1.17)$$

which means the next state S_N will only depend on the current existing state S_{N-1} and the motion or transition X_N from the existing state to the next state.

When the step length obeys the Levy distribution, such a random walk is called a Levy flight or Levy walk. If the step length is a random number drawn from Gaussian distribution the random walk is called Brawnian movement [14].

A study by Reynolds and Frye [21] shows that fruit flies, or Drosophila melanogaster, explore their landscape using a series of straight flight paths punctuated by a sudden 90° turn, leading to a Levy-flight-style intermittent scale free search pattern. Subsequently, such behavior has been applied to optimization and optimal search, and preliminary results show its promising capability [3, 21].

1.5 Chaos

Chaos means a condition or place of great disorder or confusion [23]. The first paper constructing the main principles of chaos was authored by Edward Lorenz [24]. Chaotic systems are deterministic systems showing irregular, or even random, behavior and sensitive dependence to initial conditions (SDIC). SDIC means that small errors in initial conditions lead to totally different solutions [24]. Chaos is one of the most popular phenomenan that exist in nonlinear systems, whose action is complex and similar to that of randomness [25]. A definition that is much related to optimization concepts is set in [26] as traveling particles within a limited range occurring in a deterministic nonlinear dynamic system.

Chaos theory studies the behavior of systems that follow deterministic laws but appear random and unpredictable, or we can say a dynamical system that has a sensitive dependence on its initial conditions; small changes in those conditions can lead to quite different outcomes [23]. A dynamical system must satisfy the following chaotic properties, to be referred to as chaotic [23]:

- It must be sensitive to initial conditions.

- It must be topologically mixing and its periodic orbits must be dense.

- It is ergodic and stochastically intrinsic.

Chaotic variables can go through all states in certain ranges according to their own regularity without repetition [25]. Due to the ergodic and dynamic properties of chaos variables, chaos search is more capable of hill-climbing and escaping from local optima than random search and thus has been applied to the optimization [25]. It is widely recognized that chaos is a fundamental mode of motion underlying almost natural phenomena. Given a cost function a chaotic dynamic system can reach the global optimum with high probability [25]. In mathematics, a chaotic map is a map that exhibits some sort of chaotic behavior [23].

Chaos was introduced into the optimization to accelerate the optimum seeking [25]. The chaotic map can be helpful to escape from a local minimum, and it can also improve the global/local searching capabilities [28, 29]. Some well-known chaotic maps that can be used in optimization algorithms are as follows [4]:

- *The Logistic Map*: This map is one of the simplest chaotic maps. A logistic map is a polynomial map that was first introduced by Robert May in 1976 [31]. This map is defined by:
$$x_{k+1} = ax_k(1 - x_k) \tag{1.18}$$
  where $x_k \in [0, 1]$ under the condition that $x_0 \in [0, 1]$, $0 < a \leq 4$, and k is the iteration number.

- *Lozi Map*: Lozi admits a strange attractor and the transformation is given by [32]
$$(x_{k+1}, y_{k+1}) = H(x_k, y_k) \tag{1.19}$$

$$H(x_k, y_k) = (1 + y_k - a|x_k|, bx_k) \tag{1.20}$$
  Lozi suggested the parameter values of $a = 1.7$ and $b = 0.5$.

- *Sinusoidal Map*: It is represented by [33]

$$x_{k+1} = ax_k^2 sin(\pi x_k) \tag{1.21}$$

  or its simplified form:

$$x_{k+1} = sin(\pi x_k) \tag{1.22}$$

  which generates chaotic numbers in the range (0,1).

- *Iterative Chaotic Map*: The iterative chaotic map with infinite collapses is described as [33]

$$Y_{n+1} = sin\left(\frac{\mu\pi}{Y_n}\right) \ \mu \in (0,1) \tag{1.23}$$

- *Circle Map*: The circle map is expressed as [33]

$$Y_{n+1} = Y_n + \alpha - \frac{\beta}{2\pi}sin(2\pi Y_n)mod1 \tag{1.24}$$

- *Chebyshev Map*: The family of chebyshev map is written as the following equation [33]

$$Y_{n+1} = cos(k.cos^{-1}(Y_n)) \ Y \in (-1,1) \tag{1.25}$$

$$Y_{n+1} = \mu Y_k^2 sin(\pi Y_n) \tag{1.26}$$

- *Tent Map*: This map resembles the logistic map due to its topologically conjugate behavior. The tent map can display a range of dynamical behaviors from predictable to chaotic, depending on the value of its multiplier, where the equations are 1.20 and 1.27:

$$x_{k+1} = G(x_k) \tag{1.27}$$

$$G(x) = \begin{cases} \dfrac{x}{0.7}, & x < 0.7 \\ \dfrac{1}{0.3}x(1-x) & otherwise \end{cases} \tag{1.28}$$

- *Gauss Map*: This map is a one-dimensional, real value map in the space domain which is discrete in the time domain. Gauss maps let us completely analyze its chaotic qualitative and quantitative features. The equations are as follows [33]:

$$x_{k+1} = G(x_k) \tag{1.29}$$

$$G(x) = \begin{cases} 0 & x = 0 \\ \frac{1}{x}mod1 & x \in (0,1) \end{cases} \tag{1.30}$$

- *Henon Map*: This map is a simplified version of the Poincare map of the Lorenz system [27]. The Henon equations are given by

$$y(t) = 1 - a.y(t-1) + z(t-1) \tag{1.31}$$

$$z(t) = b.y(t-1) \tag{1.32}$$

where a = 1.4 and b = 0.3 (the values for which the Henon map has a strange attractor).

- *Modified Henon map*:

$$y1(k) = 1 - a(sin(y1(k-1))) + by(k-1) \tag{1.33}$$

$$y(k) = y1(k-1) \tag{1.34}$$

where k is the iteration number. This map is capable of generating chaotic attractors with multifolds via a period-doubling bifurcation route to chaos [34]. This map exhibits very complicated dynamical behaviors with coexisting attractors. The optimization algorithm needs to normalize the variable $y(k)$ in the range [0,1] using the following transformation:

$$z(k) = \frac{v(k) - \alpha}{\beta - \alpha} \tag{1.35}$$

where α and $\beta = [-8.588 27.645]$.

1.6 Chapter conclusion

This introductory chapter aims at setting the environment for the whole book. This chapter started with basic definitions and concepts related to optimization and its sources of inspiration. It followed with a brief discussion on randomness as a basic building stone for stochastic optimization techniques. Moreover, a description of the principle of random/pseudo-number generation using common random distributions was included. Random walk is a common concept in optimization as well as chaos theory and chaos variables were becoming common as independent optimization tools or as a method for adapting optimizers' parameters, so a brief discussion about these concepts with some detail were discussed in this chapter.

Bibliography

[1] Slawomir Koziel and Xin-She Yang (Eds.), Computational Optimization, Methods and Algorithms, Studies in Computational Intelligence, Springer-Verlag Berlin Heidelberg, Vol. 356, 2011.

[2] Xin-She Yang, Nature-Inspired, Metaheuristic Algorithms, Luniver Press, 2010.

[3] Xin-She Yang, Zhihua Cui, Renbin Xiao, Amir Hossein Gandomi, and Mehmet Karamanoglu (Eds.), Swarm Intelligence and Bio-Inspired Computation: Theory and Applications, Elsevier, 2013.

[4] Jason Brownlee, Clever Algorithms Nature-Inspired Programming Recipes, Jason Brownlee Publisher, ISBN: 978-1-4467-8506-5, 2011.

[5] Iztok Fister Jr., Xin-She Yang, Iztok Fister, Janez Brest, and Dusan Fister, "A Brief Review of Nature-Inspired Algorithms for Optimization", *Elektrotrhniski, Vestnik*, Vol. 80(3), pp. 116-122, 2013.

[6] Xin-She Yang and Suash Deb, Cuckoo Search via Levy Flights. *The World Congress on Nature and Biologically Inspired Computing* (NaBIC 2009), pp. 210-214, 2009.

[7] X.-S. Yang, A New Metaheuristic Bat-Inspired Algorithm, in: Nature Inspired Cooperative Strategies for Optimization (NISCO 2010) (Eds. J. R. Gonzalez et al.), Studies in Computational Intelligence, Springer Berlin, 284, pp. 65-74, 2010.

[8] Dervis Karaboga, and Bahriye Basturk, "A powerful and efficient algorithm for numerical function optimization: artificial bee colony (ABC) algorithm", *Journal of Global Optimization*, Vol. 39(3), pp. 459-471, 2007.

[9] Scott Kirkpatrick, D. Gelatt Jr., and Mario P Vecchi, "Optimization by simulated annealing". *Science*, Vol. 220 (4598), pp. 671-680, 1983.

[10] Zong Woo Geem, Joong Hoon Kim, and GV Loganathan, "A new heuristic optimization algorithm: harmony search". *Simulation*, Vol. 76(2), pp. 60-68, 2001.

[11] Xin-She Yang, Flower pollination algorithm for global optimization, in: "Unconventional Computation and Natural Computation 2012", Lecture Notes in *Computer Science*, Vol. 7445, pp. 240-249, 2012.

[12] Kenichi Tamura and Keiichiro Yasuda, "Spiral dynamics inspired optimization". *Journal of Advanced Computational Intelligence and Intelligent Informatics*, Vol. 15(8), pp. 1116-1122, 2011.

[13] Hadi Eskandar, Ali Sadollah, Ardeshir Bahreininejad, and Mohd Hamdi, "Water cycle algorithm novel metaheuristic optimization method for solving constrained engineering optimization problems". *Computers and Structures*, Vol. 110, pp. 151-166, 2012.

[14] Scott Kirkpatrick, D. Gelatt Jr., and Mario P Vecchi, "Optimization by simulated annealing". *Science*, Vol. 220(4598), pp. 671-680, 1983.

[15] Zong Woo Geem, Joong Hoon Kim, and GV Loganathan, A new heuristic optimization algorithm: harmony search. Simulation, Vol. 76(2), pp. 60-68, 2001.

[16] Esmaeil Atashpaz-Gargari and Caro Lucas. "Imperialist competitive algorithm: an algorithm for optimization inspired by imperialistic competition". IEEE Congress on Evolutionary Computation, CEC 2007, pp. 4661-4667, 2007.

[17] Ali Husseinzadeh Kashan, "League championship algorithm: a new algorithm for numerical function optimization". International Conference in Soft Computing and Pattern Recognition, SOCPAR09, pp. 43-48, 2009.

[18] Yuechun Xu, Zhihua Cui, and Jianchao Zeng, "Social emotional optimization algorithm for nonlinear constrained optimization problems", in *Swarm, Evolutionary, and Memetic Computing*, Springer, New York pp. 583-90, 2010.

[19] David B. Thomas, Wayne Luk, Philip H.W, Leong, and John D. Villasenor, "Gaussian random number generators". *ACM Computing Surveys*, Vol. 39(4), Article 11, October 2007.

[20] Christian Walck, "Handbook on statistical distributions for experimentalists". *Internal Report SUFPFY/9601 Stockholm*, 11 December 1996, 2007.

[21] A. M. Reynolds and M. A. " Frye, Free-flight odor tracking in Drosophila is consistent with an optimal intermittent scale-free search", *PLoS* One, 2, e354 (2007).

[22] I. Pavlyukevich, "Levy flights, non-local search and simulated annealing", *J. Computational Physics*, Vol. 226, pp. 1830-1844, 2007.

[23] Rashi Vohra and Brajesh Patel, "An efficient chaos-based optimization algorithm approach for cryptography". *International Journal of Communication Network Security*, Vol. 1(4), pp. 75-79, 2012.

[24] E. N. Lorenz, "Deterministic nonperiodic flow", *J. Atmos. Sci.* Vol. 20, pp. 130-141, 1963.

[25] Bingqun Ren and Weizhou Zhong, "Multi-objective optimization using chaos based PSO", *Information Technology Journal*, Vol. 10(10), pp. 1908-1916, 2011.

[26] Min-Yuan Cheng, Kuo-Yu Huang, and Hung-Ming Chen, "A particle swarm optimization based chaotic K-means evolutionary approach", 27th International Symposium on Automation and Robotics in Construction (ISARC 2010), pp. 400-409, 2010.

[27] M. Henon, "On the numerical computation of Poincar maps", *J. Physica D*, 5(2-3) (1982) 412-414.

[28] O. Tolga Altinoz, S. Gkhun Tanyer, and A. Egemen Yilmaz, "A comparative study of FuzzyPSO and ChaosPSO", *Elektrotehniki Vestnik*, Vol. 79(1-2), pp. 68-72, 2012.

[29] O. Tolga Altinoz, A. Egemen Yilmaz, and Gerhard Wilhelm Weber, "Chaos Particle Swarm Optimized PID Controller for the Inverted Pendulum System". The

2nd International Conference on Engineering Optimization, September 6-9, Lisbon, Portugal, 2010.

[30] Afrabandpey, H. "Dept. of Electrical and Comput. Eng., Isfahan Univ. of Technol. (IUT), Isfahan, Iran; Ghaffari, M.; Mirzaei, A.; Safayani, M., A novel Bat Algorithm based on chaos for optimization tasks", Iranian Conference on Intelligent Systems (ICIS), 4-6 Feb. 2014, pp. 1-6, 2014.

[31] R. M. May, "Simple mathematical models with very complicated dynamics". *Nature*, Vol. 26, pp. 459-467, 1976.

[32] R. Caponetto, L. Fortuna, S. Fazzino, and M. Gabriella, "Chaotic sequences to improve the performance of evolutionary algorithms". IEEE Transaction on Evolutionary Computing, Vol. 7, pp. 289-304, 2003.

[33] Osama Abdel-Raouf, Mohamed Abdel-Baset, and Ibrahim El-henawy, "An Improved chaotic bat algorithm for solving integer programming problems". *I.J. Modern Education and Computer Science*, Vol. 8, pp. 18-24, 2014.

[34] E. Zeraoulia, J. C. Sprott. "A Two-dimensional Discrete Mapping with C^∞ Multifold Chaotic Attractors". *Electronic Journal of Theoretical Physics* 5, No. 17 (2008) 107-120.

2 Bat Algorithm (BA)

2.1 Bat algorithm (BA)

2.1.1 Behavior of bats

Bats are fascinating animals in that they are the only mammals with wings and they also have advanced capability of echolocation [1]. Most bats use echolocation to a certain degree; among all the species, microbats are a famous example, as microbats use echolocation extensively while megabats do not. Most microbats are insectivores. Microbats use a type of sonar, called echolocation, to detect prey, avoid obstacles, and locate their roosting crevices in the dark. These bats emit a very loud sound pulse and listen for the echo that bounces back from the surrounding objects. Their pulses vary in properties and can be correlated with their hunting strategies, depending on the species. Most bats use short, frequency-modulated signals to sweep through about an octave, while others more often use constant-frequency signals for echolocation [1]. Their signal bandwidth varies depend on the species, and often increases by using more harmonics. When hunting for prey, the rate of pulse emission can be sped up to about 200 pulses per second when they fly near their prey. Such short sound bursts imply the fantastic ability of the signal processing power of bats. The loudness also varies, from the loudest when searching for prey to a quieter base when homing toward the prey. The traveling range of such short pulses is typically a few meters, depending on the actual frequencies. Microbats can manage to avoid obstacles as small as thin human hairs. Microbats use the time delay from the emission and detection of the echo, the time difference between their two ears, and the loudness variations of the echoes to build up three-dimensional scenario of the surroundings. They can detect the distance and orientation of the target, the type of prey, and even the moving speed of the prey such as small insects. Indeed, studies suggested that bats seem to be able to discriminate targets by the variations of the Doppler effect induced by the wing-flutter rates of the target insects [1].

2.1.2 Bat algorithm

Three idealized rules can be formulated for simplicity to describe the behavior of microbats [1]:

- All bats use echolocation to sense distance, and they also know the difference between food/prey and background barriers in some magical way.

- Bats fly randomly with velocity v_i at position x_i with a fixed frequency f_{min}, varying wavelength λ and loudness A_0 to search for prey. They can automatically adjust the wavelength (or frequency) of their emitted pulses and adjust the rate of pulse emission depending on the proximity of their target.

- Although the loudness can vary in many ways, we assume that the loudness varies from a large (positive) A_0 to a minimum constant value A_{min}.

Virtual bats adjust their position according to equations 2.1, 2.2, and 2.3.

$$F_i = F_{min} + (F_{max} - F_{min})\beta \tag{2.1}$$

$$V_i^t = V_i^{t-1} + (X_i^t - X^*)F_i \tag{2.2}$$

$$X_i^t = X_i^{t-1} + V_i^t \tag{2.3}$$

where β is a random vector in the range [0,1] drawn from uniform distribution, X^* is the current global best location, F_{min} and F_{max} represent the minimum and maximum frequency need depending on the problem and V_i represents the velocity vector.

Probabilistically a local search is to be performed using a random walk as in equation 2.4.

$$X_{new} = X_{old} + \epsilon A^t \tag{2.4}$$

where A^t is the average loudness of all bats at this time and ϵ is a random number uniformly drawn from normal distribution in the range [-1 1]. It is worth mentioning that this equation can be viewed as some form of local searching. Local searching should be motivated more at the end of optimization so this updating mechanism is applied in relation to bat's pulse rate where the pulse rate decreases by time; see equation 2.5, giving more chance for local searching. It is also worth mentioning that the search locality rate should be increased as optimization grows, which means that much intensive local search is applied at the final iterations of optimization. Thus, the parameter ϵ should be decreased as optimization grows by any form of decrement or at least be static.

r_i controls the application of the local search as shown on the algorithm and is updated using equation 2.5.

$$r_i^{t+1} = r_i^0[1 - exp(-\gamma t)] \tag{2.5}$$

where r_i^0 is the initial pulse emission rate and is a constant greater than 0.

The updating of the loudness is performed using equation 2.6.

$$A_i^{t+1} = \alpha A^t \tag{2.6}$$

where α is a constant selected experimentally.

The bat's amplitude parameter is used to control the keeping/discarding of bat's new better bat's positions. If a better position is found by a given bat, it should stochastically change its current position to this new position in relation to current bat's sound amplitude.

The algorithm describing the the virtual bat algorithm is outlined in the algorithm 1.

---

input : Q frequency band
$\quad\quad\quad$ f_i pulse frequency
$\quad\quad\quad$ r_i pulse emission rate
$\quad\quad\quad$ A_i Loudness
$\quad\quad\quad$ ϵ sound amplitude control parameter
$\quad\quad\quad$ γ pulse rate control parameter
output: Optimal Bat position and the corresponding fitness

Initialize a population of n bat' positions at random
Find the best solution based on fitness; g^* ;
while *Stopping criteria not met* **do**
$\quad$ **foreach** bat_i **do**
$\quad$ Generate new solution by adjusting frequency(x_i^{new}); see equations 2.1,2.2,2.3
$\quad$ **if** $rand > r_i$ **then**
$\quad\quad$ Perform local search around global best $best$ (x_i^{new}); equation (refeq.Bat4a)
$\quad$ **end**
$\quad$ **if** $rand < A_i$ *and fitness of new solution is better than original one* **then**
$\quad\quad$ Update the position of bat_i to bex_i^{new}
$\quad\quad$ Increase emission rate r_i;2.5
$\quad\quad$ Decrease Loudness A_i;2.6
$\quad$ **end**
$\quad$ Update the best solution
end

Algorithm 1: Pseudo code for bat algorithm

2.2 BA variants

2.2.1 Discrete bat algorithm

Since standard BA is a continuous optimization algorithm, the standard continuous encoding scheme of BA cannot be used to solve many combinational optimization problems [2]. A discrete version of the standard bat optimization algorithm was proposed in [2]. Some neighborhood search methods are used to enhance the quality of the discrete solution, namely, *swap*, *insert inverse*, and *crossover*. The details of these neighborhoods are as follows:

Swap: Choose two different positions from a solution permutation randomly and swap them.

Insert: Choose two different positions from a solution permutation randomly and insert the back one before the front.

Inverse: Inverse the subsequence between two different random positions of a solution permutation.

Crossover: Choose a subsequence in a random interval from another random job permutation and replace the corresponding part of the subsequence.

The proposed discrete algorithm applies these set operators on the whole swarm iteratively given a *best* solution and hence it can update the velocity and the global solution.

2.2.2 Binary bat algorithm (BBA)

In the standard bat algorithm each bat moves in the search space toward continuous-valued positions. In some cases such as features selection the search space is modeled as an n-dimensional boolean lattice, in which the bat moves across the corners of a hypercube [3]. Since the problem is to select or not a given feature, the bat's position is then represented by binary vectors. Thus, the BBA is proposed. Sigmoidal function is common for squashing continuous values into extremes; see equation 2.7. The bat velocity will be calculated as

$$S(v_i^j) = \frac{1}{1 + \exp(-v_i^j)} \tag{2.7}$$

and the bat position is calculated as:

$$x_i^j = \begin{cases} 1 & if\, S(v_i^j) > \sigma \\ 0 & otherwise \end{cases} \tag{2.8}$$

where $\sigma\ U(0,1)$ so the position result is a 0, 1 binary.

2.2.3 Chaotic bat algorithm (CBA)

The classical BA over some benchmark functions shows that premature and/or false convergence are the most serious weaknesses of the classical BA [4]. This is mainly due to the random initialization of the parameters. In the random initialization process, the produced sequences are not well distributed and this leads to different results in different executions [4].

To deal with these problems the initialization process has to be done in a way that initial sequences are well distributed [4]. Chaotic sequences (when they are far from their strange attractors) can be a very good approach to achieve well-distributed sequences for dealing with premature and false convergence in the algorithm. To improve the performance of the classical BA, chaotic sequences are used to initialize *frequencies, loudness* and *pulse emission rates*.

Gauss map outperforms other chaotic series in enhancing convergence rate [4]. In the implementation in [4] Gauss sequences are used to initialize frequencies and loudness; see equations 2.9 and 2.10.

$$x_{k+1} = G(x_k) \tag{2.9}$$

$$G(x) = \begin{cases} 0 & x = 0 \\ \frac{1}{x} mod1 & x \in (0, 1) \end{cases} \tag{2.10}$$

This map is a one-dimensional, real value map in the space domain which is discrete in the time domain.

Tent map is used to initialize pulse emission rate; see equations 2.11 and 2.12.

$$x_{k+1} = G(x_k) \tag{2.11}$$

$$G(x) = \begin{cases} \frac{x}{0.7}, & x < 0.7 \\ \frac{1}{0.3}x(1-x) & otherwise \end{cases} \tag{2.12}$$

2.2.4 Parallel bat algorithm

In [5] a parallelized version of the standard bat algorithm was proposed. The proposed model divides the whole swarm of bats into disjoint independent subgroups, each with N bats. Each subgroup contains its own global best and its own agents. At specified time steps the information is exchanged among the subgroups. The exchange of exploration information is performed using the best of the global best solutions over all the subgroups. The obtained best of best solutions is used to replace a fraction of worst solutions in each subgroup after applying the mutation operator on it.

2.2.5 BA for constrained problems

A version of the standard bat algorithm was proposed in [6] for handling constrained optimization problems. Most design optimization problems in the real world are highly nonlinear, involving many different design variables under complex constraints. These constraints can be written either as simple bounds such as the ranges of material properties, or more often as nonlinear relationships. The nonlinearity in the objective function often results in multi-modal response landscape, while such nonlinearity in the constraints leads to the complex search domains [6]. Penalization is used in that proposed work to adapt the fitness function to be as follows:

$$\prod(x, \mu_i, \nu_j) = f(x) + \sum_{i=1}^{M} \mu_i \phi_i^2(x) + \sum_{j=1}^{M} \nu_j \psi_j^2(x) \tag{2.13}$$

where $1 \leq \mu_i$ and $0 \leq \nu_i$, which should be large enough depending on the solution quality needed. ϕ_i represents the equality constraints while ψ_j represents the inequality constraints.

By defining the new modified fitness function the target is to minimize it using the standard bat algorithm.

2.2.6 BA with Lèvy distribution

Lèvy flights are Markov processes, which differ from regular Brownian motion. Extremely long jumps may occur, and typical trajectories are self-similar, on all scales showing clusters of shorter jumps interspersed by long excursions. Lèvy flight has the following properties [7]:

- Stability: distribution of the sum of independent identically distributed stable random variables equal to distribution of each variable.

- Power law asymptotics (heavy tails).

- Generalized central limit theorem: The central limit theorem states that the sum of a number of independent and identically distributed (i.i.d.) random variables with finite variances will tend to a normal distribution as the number of variables grows.

- Infinite variance with an infinite mean value allowing for much variation.

Due to these properties the Lèvy statistics provide a framework for the description of many natural phenomena in physical, chemical, biological, and economical systems from a general common point of view [7]. Lèvy is used to enhance the local search for BA as in equation 2.14.

$$x_i^t = x_i^{t-1} + \mu sign(rand - 0.5) \oplus L\grave{e}vy \qquad (2.14)$$

where μ is a random number drawn from uniform distribution in the range from 0 to 1. $sign\oplus$ means entrywise multiplications, rand $\in [0, 1]$, and random step length that obeys Lèvy distribution.

2.2.7 Chaotic bat with Lèvy distribution

The track of chaotic variable can travel ergodically over the whole search space [8]. The chaotic variable has special characters, i.e., ergodicity, pseudo-randomness, and irregularity. The Lèvy flight process is a random walk that is characterized by a series of instantaneous jumps chosen from a probability density function that has a power law tail [8]. This process represents the optimum random search pattern and is frequently found in nature. In [8] the chaotic Lèvy flight for the improved bat algorithm is proposed. The well-known logistic map which exhibits the sensitive dependence on initial conditions is employed to generate the chaotic sequence c_s for the parameter in Lèvy flight; scaling value in bat local search. So the local searching of bat algorithm is defined as

$$x_{new} = x_{old} + c_s \bigoplus L\grave{e}vy(\lambda) \qquad (2.15)$$

where c_s is determined using logistic map and $L\grave{e}vy(\lambda)$ is a random number drawn from Lèvy distribution.

A chaotic version of the BA was proposed in [8] for solving integer programming problems. Different chaotic maps are used to update both the frequency and velocity of the bat at individual iterations as follows:

$$f_i = f_{min} + (f_{max} - f_{min})S_i \qquad (2.16)$$

where S_i is a chaos map and f_{max}, f_{min} are the bat's frequency limits.

The velocity is updated as follows:

$$v_i^t = v_i^{t-1} + (x_i^{t-1} - best)f_i * S_i \qquad (2.17)$$

where all variables are as mentioned in the standard BA and S_i is a given chaotic map. The used chaos maps are:

1. Logistic map

2. The Sine map

3. Iterative chaotic map

4. Circle map

5. Chebyshev map

6. Sinusoidal map

7. Gauss map

8. Sinus map

9. Dyadic map

10. Singer map

11. Tent map

Another variant of chaotic BA was proposed in [8]. The approach is based on the substitution of the random number generator (RNG) with chaotic sequences for parameter initialization. Simulation results on some mathematical benchmark functions demonstrate the validity of proposed algorithm, in which the chaotic bat algorithm (CBA) outperforms the classical BA.

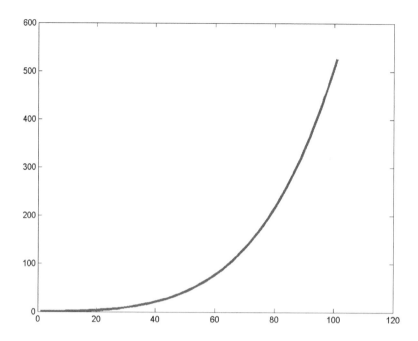

Figure 2.1: The α parameter value on each iteration (for 100 iteration)

2.2.8 Self-adaptive BA

The main idea behind self-adaptive BA is to escape from local optima as well as to avoid pre-mature convergence [9]. The α parameter controls the decrement value which affects the local search of the bat. An adaptation is proposed using equation 2.18.

$$\alpha^{new} = (1/2Iter)^{1/Iter}\alpha^{old} \tag{2.18}$$

where $Iter$ is the iteration number.

The different values for α are drawn according to equation 2.18 in the first 100 iterations in Figure 2.1. From the figure we can see that the change in amplitude of the bat sound is very small in the beginning of iteration, which allows the bat to greatly explore the search space, while in late iteration the amplitude should drop quickly to allow for much local searching.

The main problem faced in [10] is the insufficiency at exploration that exists in the BA. It can easily get trapped in local minimum on most of the multi-modal test functions. In standard BA exploration and exploitation are controlled by pulse emission rate r, and this factor increases as iteration proceeds. The algorithm gradually loses exploitation capability as iteration proceeds. To avoid this problem, exploitation capability of BA is improved by inserting linear decreasing inertia weight factor. Inertia weight is linearly decreased and hence the effect of previous velocity gradually decreases. Thus exploitation rate of BA gradually increases as iterations proceed. Equation 2.19 outlines the updating for the inertia factor.

$$w_i ter = \frac{iter_{max} - iter}{iter_{max}}(w_{max} - w_{min}) + w_{min} \tag{2.19}$$

where $iter, iter_{max}$ w_{min}, w_{max} are current iteration, maximum number of iterations, minimum value for inertia weight, and maximum inertia weight in order.

Another modification is based on changing the frequency per dimension rather than using the same frequency for all dimensions in the basic BA. Equation 2.20 outlines the calculation of the frequency value for every dimension for bat i.

$$f_j = f_{min} + \frac{\sqrt{(min(diff) - diff_j)^2}}{range}(f_{max} - f_{min}) \tag{2.20}$$

where f_j is the frequency for dimension j, f_{max} and f_{min} are the frequency minimum and maximum, and $diff_j$ is the difference in dimension j between the *best* solution and the Bat position i.

The step size updating for the standard BA algorithm is proposed in [11]. Proper tuning of this parameter reduces the number of iterations (hence the computational time). In addition, this improvement can provide easy adjustment of the bat algorithm for discrete optimization problems. Dynamic scale factor parameter is formulated as

$$\theta^{iter} = \theta_{max} \exp(\frac{\ln(\frac{\theta_{min}}{\theta_{max}})}{\theta_{max}}.iter) \tag{2.21}$$

and, hence, the local search of the BA is updated as follows:

$$x_{new} = x_{old} + \varepsilon\theta^{iter} \tag{2.22}$$

where θ^{iter} is the step size at iteration $iter$, $\theta_{min}, \theta_{max}$ are the minimum and maximum step size in order, and ε is a suitable random number.

2.2.9 Adaptive bat algorithm

A simplified bat position updating equation is proposed in [12]. The improved algorithm is more concise and clear. The search process is only controlled by the position vector; each bat updates its location which is only dependent on bat individual global optimal location p_{gd} and random location pad in the current. p_{ad} can increase the variability of the bat position; to some extent it increased the diversity of the bats; see equation 2.23.

$$x_i^{t+1} = wx_i^t + f_{1t}(p_{ad} - x_i^t) + f_{2t}(p_{gd} - x_i^t) \tag{2.23}$$

$$p_{ad} = 2 * x_i^t * rand \tag{2.24}$$

where t is the current iteration number, w is the inertia weight, f_{1t}, f_{2t} are the frequency, x_t^i is i-th bat location in the current iteration, and p_{gd} is global optimum location in the current iteration, p_{ad} is a bat individual random position in the current iteration.

In order to make the global search ability and local search ability to be balanced, we can achieve their goals through the inertia weight adjustment. The strategy of inertia weight w commonly is used with linear decreasing. Large values for inertia weight are conducive to the global search, but the algorithm overhead is large and search efficiency is low. In the latter part we can get smaller values conducive to accelerate the convergence of the algorithm, but the algorithm is easy to fall into local optimum and lacks the ability to improve the solution. So the algorithm will set inertia weight as random numbers which obey normal distribution. That is not only conducive to maintain the diversity of the population, but also improves the performance of the global search in the latter part of the algorithm; refer to equation 2.25.

$$w = \mu_{min} + (\mu_{max} - \mu_{min}) * rand + \sigma(0.5 randn + 0.5 poissrnd) \qquad (2.25)$$

where $rand, randn, poissrnd$ are random numbers drawn from uniform, normal, and poisson distributions, μ_{max}, μ_{min} are the maximum and minimum value for inertia, and σ is the variance of the random weight.

According to the loudness and pulse emission rate updated with the iterative process, we consider that the frequency can be adaptive while the number of iterations increases. We will take a strategy that f will be divided into f_1 and f_2. In the early search stage when f_1 is low value and f_2 is high, the bats would be closer to the average position of the bat colony and less close to the optimal position, which is conducive to the bat colony, and strengthens the capability of the global search. In the latter search stage when f_1 is high value and f_2 is low, the bats would be closer to the optimal position and less close to the average position of the bat colony, which is of rapid convergence to global optimal solution. The updating for f_1, f_2 is as in equation 2.24.

$$f_1(t) = 1 - \exp(-|F_{avg}(t) - F_{gbest(t)}|) \qquad (2.26)$$

$$f_2(t) = 1 - f_1(t) \qquad (2.27)$$

$$F_{avg}(t) = \frac{1}{N} \sum_{t=1}^{N} (f(x_i(t))) \qquad (2.28)$$

where $f_1(t), f_2(t)$, respectively, represent the frequency, and $F_{avg}(t)$ and $F_{gbest}(t)$, respectively, represent the adaptation value of the bat colony average fitness and the optimal location in the iteration t.

2.3 Bat hybridizations

In the literature many hybridizations are proposed between firefly algorithm and other optimizers to enhance its performance in solving optimization tasks. Some of the BA problems reported in the literature are:

Table 2.1: BA hybridizations and sample references

Hybrid with	*Target*	*Sample References*
Differential evolution (DE)	To enhance local searching heuristics, and exploit the self-adaptation mechanisms	[13, 14]
Particle swarm optimization (PSO)	To ensure population diversity and replace weak agents	[15]
Cuckoo search (CS)	To enhance accuracy, convergence speed, and ensure global convergence	[16]
Simulated annealing (SA)	To exploit global searching capability of SA and speed up convergence rate	[17]
Harmony search (HS)	To hire the pitch adjustment into mutation operators to ensure global convergence	[19]
Artificial bee colony (ABC)	To enhance population diversity and enhance global convergence	[18]

- BA has low convergence accuracy, slow convergence velocity, and easily falls into local optima.

- It cannot perform a global search well.

- It suffers from lack of incorporated domain-specific knowledge of the problem to be solved.

Table 2.1 states samples of hybridizations mentioned in the literature.

2.3.1 Bat hybrid with differential evolution (DE)

The original BA algorithm suffers from a lack of incorporated domain-specific knowledge of the problem to be solved [13]. That work focuses on hybridizing the BA using novel local search heuristics that better exploit the self-adaptation mechanism of this algorithm. The standard rand/1/bin DE strategy and three other DE strategies focusing on the improvement of the current best solution were used for this purpose.

The local search is launched according to a threshold determined by the pulse rate r_i. The local search is an implementation of the operator's crossover and mutation borrowed from DE. First, the four virtual bats are selected randomly from the bat population and the random position is chosen within the virtual bat. Then, the appropriate DE strategy modifies the trial solution. The DE strategies are many and always direct the solution toward the global best solution. Equations 2.29, 2.30, 2.31, and 2.32 outline sample DE strategies used.

$$y_j = x_{r1,j} + F.(x_{r2,j} - x_{r3,j}) \tag{2.29}$$

$$y_j = x_{i,j} + F.best_j(x_{r1,j} - x_{r2,j}) \qquad (2.30)$$

$$y_j = best_j + F.(x_{r1,j} + x_{r2,j} - x_{r3,j} - x_{r4,j}) \qquad (2.31)$$

$$y_j = best_j + F.(x_{r1,j} - x_{r2,j}) \qquad (2.32)$$

where F is a constant scaling factor always in $[0.1, 1]$, *best* is the global best solution, and x_r is a randomly selected virtual bat.

Another hybridization is proposed in [14] where the DE strategy DE/rand/1/bin is used to replace the local search using normal distribution in the original BA algorithm. DE/rand/1/bin denotes that the base vector is randomly selected, 1 vector difference is added to it, and the number of modified parameters in mutation vector follows binomial distribution.

2.3.2 Bat hybrid with particle swarm optimization

A hybrid particle swarm optimization with bat algorithm (hybrid PSO-BA) is designed based on original PSO and bat algorithm in [15]. Each algorithm evolves by optimization independently, i.e., the PSO has its own individuals and the better solution to replace the worst artificial bats of BA. In contrast, the better artificial bats of BA are to replace the poorer individuals of PSO after running some fixed iterations. Figure 2.2 outlines the communication strategy used to exchange solutions between PSO and BA agents.

2.3.3 Bat hybrid with cuckoo search

In order to solve the problems of bat algorithm including low convergence accuracy, slow convergence velocity, and easily falling into local optimization, this paper presents an improved bat algorithm [16]. Based on basic CS algorithm, it draws the bat algorithm into the basic CS algorithm, and makes their respective advantages together in a hybrid optimization algorithm (BACS), bat algorithm, and cuckoo optimization algorithm.

Based on the basic algorithm of the cuckoo, the bird's nest location x_i^t does not access to the next iteration $(t + 1)$ directly after the evolution, but access to BA algorithm continues to update the bird's nest position through the dynamic transformation strategy. First of all, comparing a uniformly distributed random number with the launch and pulse rate, if it meets the conditions, a random disturbance is carried out for the optimal position of the bird's nest with current, and gains a new bird's nest. Then it carries on the cross-border process and evaluates the corresponding of the fitness value.

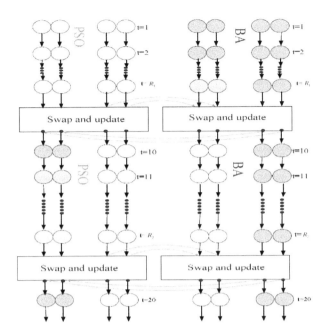

Figure 2.2: Solution exchange between bat algorithm and particle swarm optimization

Make a comparison between loudness and uniformly distributed random number, and update the bird's nest position in the basic cuckoo algorithm by applying the new operators if conditions are met. Update the launch of loudness and pulse rate as well. Finally evaluate the fitness values of the bird's nest, and find out the current optimal position of the bird's nest and the optimal value, and enter the next iteration; continue to search and update the position through adopting the basic cuckoo algorithm. It solves the balance of global search and local search well in the algorithm, thus improving the convergence of the algorithm, and avoiding it falling into local optimum and getting the global optimal solution. And it enhances the ability of local optimization and convergent precision at the same time.

2.3.4 Bat hybrid with simulated annealing

A simulated annealing Gaussian bat algorithm (SAGBA) for global optimization was proposed in [17]. The proposed algorithm not only inherits the simplicity and efficiency of the standard BA with a capability of searching for global optimality but also speeds up the global convergence rate.

The basic idea and procedure can be summarized as two key steps: Once an initial population is generated, the best solutions are replaced by new solutions generated by using simulated annealing (SA), followed by the standard updating equations of BA. Then, the Gaussian perturbations are used to perturb the locations/solutions to generate a set of new solutions.

The basic steps for the SAGBA are as follows:

1. Initialize the bat positions, velocities, frequency ranges, pulse rates, and the loudness as in the standard BA.

2. Evaluate the fitness of each bat, store the current position and the fitness of each bat, and store the optimal solution of all individuals in the *pbest*.

3. Determine the initial temperature as in the equation below:

$$t_0 = \frac{f(pbest)}{\ln 5} \tag{2.33}$$

4. According to the following formula to determine the adaptation value of each individual in the current temperature,

$$TF(x_i) = \frac{e^{-(f(x_i)-f(pbest))/t}}{\sum_{j=1}^{N} e^{-(f(x_j)-f(pbest))/t}} \tag{2.34}$$

where t is the current temperature.

5. Use roulette strategy to determine an alternative value *pbest* and then update the velocity and position as in the standard BA.

6. Calculate the new objective or fitness value of each bat, and update the positions/solution if it is better. Then, carry out Gaussian perturbations, and compare the position before and after Gaussian perturbations to find the optimal position *pbest* and its corresponding optimal value. Gaussian perturbations can be simply implemented as

$$X^{t+1} = X^t + a\varepsilon \tag{2.35}$$

where $\varepsilon \sim N(0,1)$ and a is a scaling factor.

7. Update the cooling schedule as

$$t_{k+1} = \lambda t_k \tag{2.36}$$

where where λ in $[0.5, 1]$ is a cooling schedule parameter.

8. Check for stopping and repeat at step 4.

2.3.5 Bat hybrid with harmony search

The standard BA algorithm is adept at exploiting the search space, but at times it may trap into some local optima, so that it cannot perform a global search well [19]. A hybrid meta-heuristic algorithm induces the pitch adjustment operation in HS as a mutation operator into a bat algorithm, the so-called harmony search/bat algorithm(HS/BA). The difference between HS/BA and BA is that the mutation operator is used to improve the original BA generating a new solution for each bat. In this way, this method can explore the new search space by the mutation of the HS algorithm and exploit the population information with BA, and therefore can avoid trapping into local optima in BA [19]. In the HS/BA a local search is performed either using the standard bat local search or stochastically using the harmony search equation as follows:

$$x_{new} = x_{old} + bw(2\varepsilon - 1) \qquad (2.37)$$

where ε is a random real number drawn from a uniform distribution [0, 1] and BW is the bandwidth controlling local range of pitch adjustment. Pitch adjustment operation in HS serves as a mutation operator updating the new solution to increase diversity of the population to improve the search efficiency.

2.3.6 Bat hybrid with artificial bee colony

The idea is based on replacing the weaker individuals according to a fitness evaluation of one algorithm with stronger individuals from an other algorithm in parallel processing for swarm intelligent algorithms [18]. Several groups in a parallel structure are created from dividing the population into subpopulations to construct the parallel processing algorithm. Each of the subpopulations evolves independently in regular iterations. They only exchange information between populations when the communication strategy is triggered. It results in taking advantage of the individual strengths of each type of algorithm, replacing the weaker individuals with the better one from the other, and reducing the population size for each population and the benefit of cooperation is achieved.

The hybrid BA-ABC is designed based on original BA optimization and artificial bee colony optimization algorithm. Each algorithm evolves by optimization independently, i.e., the BA has its own bats and near best solution to replace artificial agents of ABC worst and not near best solution. In contrast, the artificial agents better than ABC are to replace the poorer bats of BA after running R_i iterations. The total iteration contains R times of communication, where $R = R_1, 2R_1, 3R_1, \ldots$. The bats in BA don't know the existence of artificial bees of ABC in the solution space [18]. Figure 2.3 outlines an abstracted view of the communication between BA and ABC [18].

2.4 Bat algorithm in real world applications

This section presents sample applications for the bat algorithm (BA). BA was successfully applied in many disciplines in decision support and making, image processing domain, and engineering. The main focus is on identifying the fitness function and the optimization variables used in individual applications. Table 2.2 summarizes sample applications and their corresponding objective(s).

BA is used in [20] for feature reduction. The BA is used to search the space of feature combination to find a feature set maximizing the rough-set classifier. It is used with naive

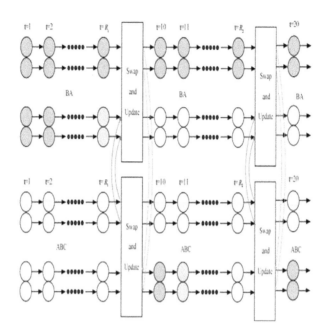

Figure 2.3: The diagram of hybrid BA-ABC with communication strategy [18]

Bayes theory for feature selection in [21]. The velocity in the standard BA is represented as a positive integer number. Velocity suggests the number of bat attributes that should change at a certain moment of time. The difference between bat position and best solution in velocity calculation is represented as the difference between feature size of the given solution and the best solution. Bat position is represented as the binary string representing the selected/unselected features. Loudness is represented as the change in number of features at a certain time during local search around the global best bat, as well as a local search around the i-th bat. The fitness function used is a double objective one that incorporates both classification accuracy and feature reduction; refer to equation 2.38.

$$f = 0.9 \frac{TP + TN}{TP + TN + FP + FN} + 0.1 \frac{TF - SF}{TF} \tag{2.38}$$

where TP, FP, TN, FN are the classifier true positives, classifier false positive, classifier true negatives, and classifier false negative in order and TF, SF are the total number of features and the size of selected features in order.

Predicting efforts for the software testing phase is a complex task. There are many factors that affect test effort estimation, such as productivity of the test team, testing strategy, size and complexity of the system, technical factors, expected quality, and others [22]. There is a strong need to optimize the test effort estimation. BA is used to estimate the test effort. The proposed model is used to optimize the effort by iteratively improving the solutions. The amount of deviation between the expected and the actual effort is to be minimized so that the prediction system will be suitable.

Table 2.2: CS applications and the corresponding fitness function

Application Name	Fitness Function	Sample References
Feature selection and ranking	Maximize classification accuracy and minimize number of selected features	[20, 21]
Prediction of software test effort	The fitness is the amount of deviation between the predicted and actual test effort	[22]
Resource scheduling in cloud environment	Objective is to minimize the makespan and time	[23]
Localization of nodes in wireless sensor network	Fitness is estimated as the amount of deviation between actual and predicted positions of some known-position nodes	[24]
Cluster head selection in wireless sensor network	Goal is to maximize network lifetime by minimizing routing cost	[25]
Economic load dispatch (ELD)	Target is minimizing the total fuel cost while enhancing security and stability of power systems	[26]
Economic load dispatch	The objective of the economic dispatch problem is minimization of operating cost	[?]
Design of brushless DC wheel motor	Maximizing the efficiency of the DC wheel motor	[27]
Microstrip coupler design	Minimizing microstrip coupler design equations	[28]
Training of neural network	Target is the amount of deviation between the expected and the actual network output	[29]
Training of functional Link Artificial Neural Network (FLANN) classifier	Objective is to minimize the classification error rate for the network	[30]
To find optimal parameter values for support vector machine (SVM)	Error between predicted and actual output of the SVM	[31]
Image thresholding	maximizing Kapur's entropy	[32]
Image thresholding	Maximizing Otsus between class variance function	[33]
Image thresholding	Minimizing variance inside individual clusters	[34]
Image thresholding	Maximize the correlation between data points in same cluster	[35, 36]
Multi-level image thresholding	Maximize entropy inside individual clusters	[37]

BA is used in [23] for resource scheduling in cloud computing. Resource scheduling is a complicated process in cloud computing because of the heterogamous nature of the cloud and multiple copies of the same task is given in multiple computers. The BA proves good accuracy and efficiency in its application. The fitness for evaluating the bat solution is composed of two components, namely, makespan and cost. The makespan is the total execution time while the cost is measured in dollars and it is decided per job.

In wireless sensor network (WSN) the information gathered by the microsensors will be meaningless unless the location from where the information is obtained is known. This makes localization capabilities highly desirable in sensor networks [24]. Using the global position system is inefficient and costly. So, instead of requiring every node to have GPS installed, all localization methods assume only a few nodes be equipped with GPS hardware. These nodes are often called anchor nodes and they know their positions. Other normal sensors can communicate with a few nearby sensors and estimate distances between them using some localization algorithm and then derive their positions based on the distances. BA is employed to localize nodes based on their estimated distance to each other [24].

BA is used in [25] in WSN for cluster head selection. The target of the BA is to divide the nodes of the network into a set of disjoint clusters and further a node inside each cluster is used as a cluster head.

Power system operation involves some kind of optimization for ensuring economy, security, and stability. Economic load dispatch (ELD) is one such optimization problem and it is applied for minimizing the total fuel cost. Optimizing the fuel cost is done by properly setting the real power generation from the generators in a power system [26]. BA is used in optimization of ELD in [26].

ELD is an optimization problem for scheduling generator outputs to satisfy the total load demand at the least possible operating cost. An ELD problem is often formulated as a quadratic equation. The ELD problem is in reality a nonconvex optimization problem [26]. The objective of the economic dispatch problem is minimization of operating cost. The generator cost curves are represented by a quadratic function with a sine component. The sine component denotes the effect of steam valve operation [26]. The fitness function can be formulated as follows:

$$F_c(P_g) = \sum_{i=1}^{Ng} a_i P_i^2 + b_i P_i + c_i + |d_i sin[e_i * (p_i^{min} - p_i)]| \qquad (2.39)$$

where Ng is the number of generating units, a_i, b_i, c_i, d_i *and* e_i are the cost coefficients of the ith generating unit, and p_i is the real power output of the ith generator.

The aim of BA in [27] is to optimize the mono- and multi-objective optimization problems related to the brushless DC wheel motor problems, which has five design parameters and six constraints for the mono-objective problem and two objectives, five design parameters, and five constraints for the multi-objective version. The goal of the optimization is to maximize the efficiency of the DC wheel motor.

Microstrip couplers are part of many microwave circuits and components and their design is of interest [28]. The theoretical foundations and the models predict very close proximity to the real experimental results that are obtained, however the complexity of equations make the design a difficult task unless an optimizing tool in a CAD program is used. These equations are used in an algorithm and then the obtained results are used in a CAD program to observe whether the algorithm has found correct dimension values for the desired coupling. The target is to minimize microstrip coupler design equations [28].

BA is used to find optimal weights and biases in a momentum feed-forward neural network [29]. The target or fitness function is to minimize the network output error. In [30], a model has been proposed for classification using bat algorithm to update the weights of a Functional Link Artificial Neural Network (FLANN) classifier. The proposed model has been compared with FLANN, PSO-FLANN. The objective function is the error between the targeted output and the actual output. BA is used to find optimal parameter values for the support vector machine (SVM) [31]. The trained SVM is used to predict future trends of the stock market.

Threshold localization based on maximizing Kapur's entropy using BA is proposed in [32]. The selected threshold divides the histogram into non-overlapping regions.

Histogram based bilevel and multi-level segmentation is proposed for gray-scale images using bat algorithm (BA). The optimal thresholds are attained by maximizing Otsus between class variance function [33].

In [34] BA has been used for detection of hairline bone fracture. BA is used as a pre-processor for image enhancement and then image is passed to a self-organizing map for further clustering. BA is used to adjust threshold based on minimizing variance inside individual clusters.

In [35] BA is used to find optimal cluster centers. The distance similarity measure is the correlation between the data points and the cluster center; refer to equation 2.40.

$$r = \frac{\sum_{i=1}^{n} x_i y_i}{\sqrt{\left(\sum_{i=1}^{n} x_i\right) * \left(\sum_{i=1}^{n} y_i\right)}} \tag{2.40}$$

where x, y, n are the first vector, second vector, and number of dimensions.

The monochromatic cost function is suitable for detecting meaningful homogeneous bi-clusters based on categorical valued input matrices [36]. Given an input matrix over some fixed finite domain D of values, and integers K and L, the monochromatic bi-clustering task is to find a partition of the rows of the matrix into K groups and its columns into L groups, such that the resulting matrix blocks are as homogeneous as possible. We define the cost of a partition as the fraction of matrix entries that reside in blocks in which they are not the majority value entries. The BA iterates over all (discarding symmetries) possible patterns, and for each pattern finds the best (approximated) partition. Naturally, the pattern of the best partition will be considered, and it is easy to see that the partition minimizing the cost with respect to the optimal pattern is the optimal monochromatic solution [36].

A system for multi-level image thresholding was proposed in [37] based on the bat algorithm. The optimal thresholds are obtained by maximizing the objective function:

$$min\left(\sum_{i=0}^{k} H_i\right) \tag{2.41}$$

where k is the number of thresholds required and H_i is the entropy inside cluster i.

2.4.1 Bat algorithm for solving ill-posed geomagnetic inverse problem

The aim of the geomagnetic inverse problem is to detect the buried magnetic bodies from the measured geomagnetic field signal. In the two-dimensional space, the model domain is subdivided into two-dimensional prisms of equal sides and each prism is extended to infinity in the third direction perpendicular to the plane. Each prism has an unknown magnetic susceptibility. The problem can be expressed as

$$\min_{x} = \|Ax - b\|^2 \tag{2.42}$$

where A is the kernel matrix, which expresses the geometrical relation between each prism and each measuring point, x is the vector of unknown magnetic susceptibility of prisms, and b is the vector of measured magnetic field. To solve the above system, the BA as an evolutionary technique is a good choice.

The Kaczmarz method or the modified version of Kaczmarz algorithm for solving ill-posed systems of linear algebraic equations is based on transforming regularized normal equations to the equivalent augmented regularized normal system of equations [38]. Such algorithm can effectively solve ill-posed problems of large dimensionality. By running this method on different data samples we found that it converges to a suboptimal solution especially if the problem dimension is very high. Thus, the Kaczmarz were used as an initializer for the bat algorithm for its fast regularization, and further, the BA adapts this solution, as well as other random solutions, using BA principles. In the Kac-BA the Kaczmarz is set to run around one third of the total running time and BA is run to enhance the solution by Kaczmarz for the rest of the allowed time. For comparison purposes we compared the Kaczmarz method for regularization against a Kac-Bat regularization where both methods are allow to run for the same running time and each method is run on the same data for 10 runs to estimate statistically the capability of each method to converge regardless of the initial and random solutions used. The new hybrid method was applied to three synthetic examples using different noise levels and earth model dimensions in each case to prove its efficiency.

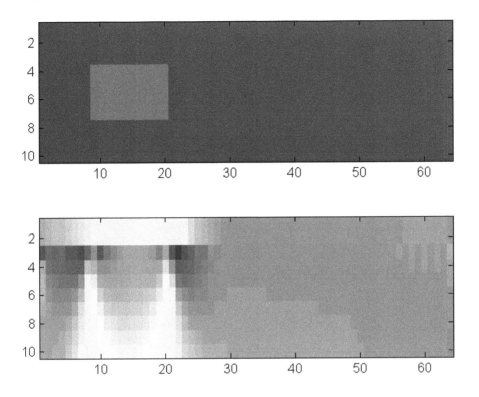

Figure 2.4: Sample synthetic data and the corresponding reconstructed image using Kaczmarz method at SNR = 100 at dimension [10 × 64]

Three synthetic data samples were used for the experiment and testing the algorithm. Each data sample is exposed to white Gaussian noise to simulate the natural field data where different noise may be observed with signal-to-noise ratio (SNR) ranges from 100 to 60 with step 10. The synthetic data were resized to different sizes to evaluate the performance of each optimizer on low and high dimensional problems. The problem dimensions were [10x64], [10x16], and [5x16].

Figure 2.4 displays sample synthetic data and the corresponding reconstructed image using Kaczmarz method at SNR = 100 at dimension [10x64]. We can see that the Kaczmarz can guess the position of the mass but with some haze boundaries and mispositioning.

Figure 2.5 displays sample synthetic data and the corresponding reconstructed image using BA Method at SNR = 100 at dimension [10x64]. We can see that the BA can localize the synthetic object in both dimensions.

Figure 2.6 displays sample synthetic data and the corresponding reconstructed image using BA method at SNR = 100 at dimension [5x16]. We can see that the BA can guess the position of the mass much better at this dimension which is, respectively, low.

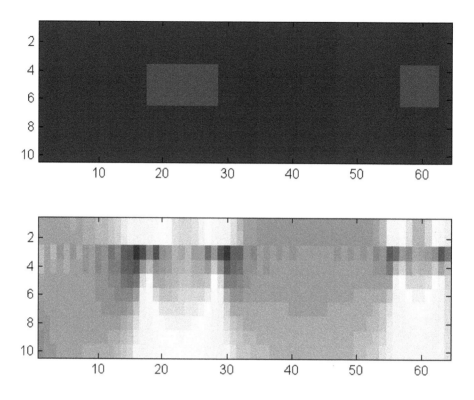

Figure 2.5: Sample synthetic data and the corresponding reconstructed image using BA method at SNR = 100 at dimension [10 × 64]

Table 2.3 outlines the performance of BA and Kaczmarz over the first synthetic data on different noise rates and on different dimensions. The output fitness value from Kaczmarz and Kac-BA are outlined as well as the mean square error between the reconstructed image and the synthetic one. We can see remarkable advance for the BA over Kaczmarz at the high dimensional data, which can be interpreted by the exploitative capabilities of BA especially at the final steps for optimization. We can also remark that noise effect on the performance of BA is minor.

At data with lower dimension we can see that Kaczmarz can reach the global solution and hence it outperforms BA. Similar results and conclusion can be derived from Tables 2.4 and 2.5 on synthetic data sample 2 and 3.

Figure 2.8: Visualizations of the result obtained by bat algorithm on Facebook social network dataset

2.4.2　Bat algorithm for the community social network detection problem

Community detection in networks has raised an important research topic in recent years. The problem of detecting communities can be modeled as an optimization problem where

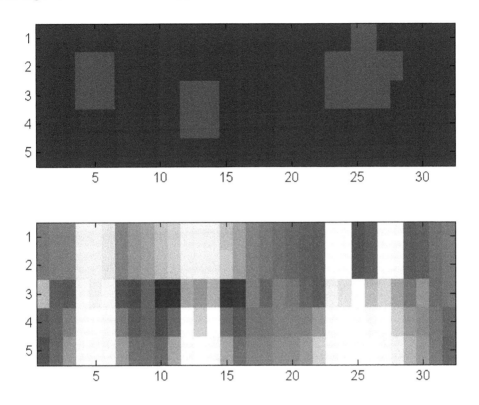

Figure 2.6: Sample synthetic data and the corresponding reconstructed image using Kaczmarz method at SNR = 100 at dimension [10 × 64]

Table 2.3: Resulting fitness value and mean square error for BA and Kaczmarz on synthetic data 1 at different SNR and different dimensions

SNR	Dim	Kaczmarz Fitness	Kac-BA Fitness	Kaczmarz MSE	BA MSE
100	640	18.56654	15.70077	0.000004	0.000004
100	160	4.364369	3.211048	0.000002	0.000002
100	80	1.82478	2.315475	0.000002	0.000004
90	640	17.80673	15.93999	0.000004	0.000004
90	160	4.254198	3.220406	0.000002	0.000002
90	80	1.839903	2.346384	0.000002	0.000004
80	640	18.13617	16.13885	0.000004	0.000004
80	160	4.39544	3.258705	0.000002	0.000002
80	80	1.881141	2.37184	0.000002	0.000004
70	640	17.72511	17.26083	0.000004	0.000004
70	160	4.397793	3.49102	0.000002	0.000002
70	80	1.964255	2.575199	0.000002	0.000004
60	640	23.32973	21.53744	0.000004	0.000004
60	160	5.456999	4.243028	0.000002	0.000002
60	80	2.312926	3.843799	0.000002	0.000004

Table 2.4: Resulting fitness value and mean square error for BA and Kaczmarz on synthetic data 2 at different SNR and Different Dimensions

SNR	Dim	Kaczmarz Fitness	Kac-BA Fitness	Kaczmarz MSE	BA MSE
100	640	20.14454	17.67157	0.000005	0.000005
100	160	4.8768	4.057205	0.000004	0.000004
100	80	3.007872	4.913887	0.000004	0.000005
90	640	19.98031	17.74035	0.000005	0.000005
90	160	5.19905	4.045526	0.000004	0.000004
90	80	3.207954	5.002001	0.000004	0.000005
80	640	20.67425	17.63531	0.000005	0.000005
80	160	5.898271	4.248078	0.000004	0.000004
80	80	3.362344	5.026152	0.000004	0.000005
70	640	21.02544	18.18722	0.000005	0.000005
70	160	5.292449	4.127389	0.000004	0.000004
70	80	3.3517	4.897991	0.000004	0.000005
60	640	21.67418	19.0649	0.000005	0.000005
60	160	6.335895	5.229592	0.000004	0.000004
60	80	4.417849	5.067848	0.000004	0.000005

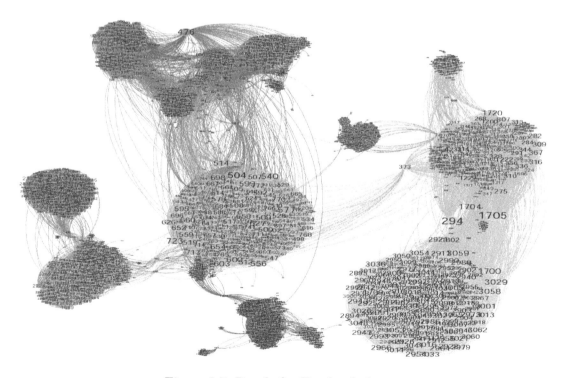

Figure 2.7: Result for Facebook dataset

Table 2.5: Resulting fitness value and mean square error for BA and Kaczmarz on synthetic data 3 at different SNR and different dimensions

SNR	*Dim*	*Kaczmarz Fitness*	*Kac-BA Fitness*	*Kaczmarz MSE*	*BA MSE*
100	640	34.29171	29.05363	0.000008	0.000009
100	160	9.053332	7.218613	0.000006	0.000006
100	80	5.408153	6.843847	0.000006	0.000008
90	640	32.52878	29.32763	0.000008	0.000009
90	160	9.36744	7.284556	0.000006	0.000006
90	80	5.210043	6.787388	0.000006	0.000008
80	640	33.22117	29.30881	0.000008	0.000009
80	160	9.684629	7.226406	0.000006	0.000006
80	80	5.115863	6.751183	0.000006	0.000008
70	640	34.45014	29.81535	0.000008	0.000009
70	160	10.22134	7.522129	0.000006	0.000006
70	80	5.157756	6.950278	0.000006	0.000008
60	640	38.6098	35.45318	0.000008	0.000009
60	160	11.88874	9.294612	0.000006	0.000006
60	80	6.157133	7.277374	0.000006	0.000008

a quality objective function that captures the intuition of a community as a set of nodes with better internal connectivity than external connectivity is selected to be optimized. In this work the bat algorithm was used as an optimization algorithm to solve the community detection problem. Bat algorithm is a new nature-inspired meta-heuristic algorithm that proved its good performance in a variety of applications. However, the algorithm performance is influenced directly by the quality function used in the optimization process. Experiments on real life networks show the ability of the bat algorithm to successfully discover an optimized community structure based on the quality function used and also demonstrate the limitations of the BA when applied to the community detection problem.

In this section we tested our algorithm on a real life social network for which a ground truth community partition is known. To compare the accuracy of the resulting community structures, we used normalized mutual information (NMI) [] to measure the similarity between the true community structures and the detected ones. Since modularity is a popular community quality measure used extensively in community detection, we used it as a quality measure for the result community structure of all other objectives.

We test the BA algorithm on Facebook dataset [?]. The data were collected from survey participants using a Facebook application [?]. The ego network consists of a user's – the ego node – friends and their connections to each other. The 10 Facebook ego networks from [?] are combined into one big network. The resulting network is an undirected network that contain 3959 nodes and 84243 edges. Despite that there is no clear community structure for the network, a ground truth structure was suggested in [?]. For the tested dataset, we applied the bat algorithm 10 restarts and calculated the NMI and modularity value of the best solution selected. This process was repeated 10 times and average NMI and average modularity are reported. The bat algorithm was applied with the following parameter values: number of BA in the population $np = 100$ and the maximum number of iterations $Max\_Iterations = 100$.

Figure 2.7 shows a visualization of the largest 12 communities from the detected community structure of the Facebook dataset. We can observe that each group shows a dense connection between nodes from the same group and sparce or low interactions between nodes from different groups, except for a few nodes that have a large edge degree across different groups such as nodes {1750, 476, and 294}, which makes group membership assignment a hard process for the algorithm.

As observed from the experimental result the bat algorithm performance is promising for small-size networks, however for large networks bat algorithm performance is degraded compared to other community detection algorithms. From our initial analysis, we found there is not much diversity in the BA swarm over the search space and the bat algorithm does not explore a large region in the search space for the following reasons:

- All the population is moving toward one position (current global best x^*). Over iterations this will lead to all BAs will moving/evolving to similar solutions. Despite that we overcome this problem using η top global best, it decreased its impact but did not eliminate it.

- There is no operator/behavior that allows the BA to escape a local optima or jump/explore new random regions in the search space. For example, in genetic algorithm there exists a mutation operation that allows such behavior that even a simple mutation in the current solution could cause a large diversity in the current population. Despite that bat algorithm performance in other applications that continue in nature is very promising, for the community detection problem (discrete case) bat algorithm performance has some limitations.

- The local search and bat difference operator are not optimal and it is not clear if they are efficient in exploring the search space. It is possible for another design to cause a significant improvement to the algorithm performance.

- Accepting criteria for new solutions has some limitation. The basic bat algorithm accepts new solutions only if it is better than the current global best. This may constrain the number of moves that a bat can perform.

2.5 Chapter conclusion

This chapter discusses the behavior of bats and its variants. Variant versions include discrete bat algorithm, binary bat algorithm, chaotic bat algorithm, parallel bat algorithm, bat for constrained problems, bat with Lèvy distribution, chaotic bat with Lèvy distribution, self-adaptive bat algorithm, and an adaptive bat algorithm. Moreover, this chapter reviews the most hybrid evolutionary bat algorithm with other optimization techniques. This chapter ends by discussion of a real-world application as a case study of bat algorithm as well as shows how bat algorithm is applied for solving ill-posed geomagnetic inverse problems and detection in the community in the social network.

Bibliography

[1] X.-S. Yang, "A New Metaheuristic Bat-Inspired Algorithm", in: *Nature Inspired Cooperative Strategies for Optimization* (NISCO 2010) (Eds. J. R. Gonzalez et al.), Studies in Computational Intelligence, Springer Berlin, 284, pp. 65-74, 2010.

[2] Qifang Luo, Yongquan Zhou, Jian Xie,Mingzhi Ma, and Liangliang Li, "Discrete bat algorithm for optimal problem of permutation flow shop scheduling", *Scientific World Journal*, Vol. 2014, Article ID 630280, 2014, http://www.hindawi.com/journals/tswj/2014/630280/.

[3] R.Y.M. Nakamura, L.A.M. Pereira, K.A. Costa, D. Rodrigues, J.P. Papa, and X. S. Yang, BBA: "A Binary Bat Algorithm for Feature Selection", 25th SIBGRAPI Conference on Graphics, Patterns and Images (SIBGRAPI), 22-25 Aug. 2012, pp. 291-297, Ouro Preto, 2012.

[4] H. Afrabandpey, M. Ghaffari, A. Mirzaei, and M. Safayani, "A novel Bat Algorithm based on chaos for optimization tasks", Iranian Conference on Intelligent Systems (ICIS), 4-6 Feb. 2014, pp. 1-6, 2014.

[5] Moonis Ali, Jeng-Shyang Pan, Shyi-Ming Chen, and Mong-Fong Horng, Modern "Advances in Applied Intelligence", 27th International Conference on Industrial Engineering and Other Applications of Applied Intelligent Systems, IEA/AIE 2014, Kaohsiung, Taiwan, June 3-6, Part I, pp. 87-95, 2014.

[6] Amir Hossein Gandomi, Xin-She Yang, Amir Hossein Alavi, and Siamak Talatahari, "Bat algorithm for constrained optimization tasks", *Neural Comput and Applic.*, Vol. 22, pp. 1239-1255, 2013.

[7] Jian Xie, Yongquan Zhou, and Huan Chen, "A novel bat algorithm based on differential operator and Levy flights trajectory", *Computational Intelligence and Neuroscience*, Vol. 2013, Hindawi Publishing Corporation, Article ID 453812, 13 pages, http://dx.doi.org/10.1155/2013/453812.

[8] Jiann-Horng Lin, Chao-Wei Chou, Chorng-Horng Yang, and Hsien-Leing Tsai, "A chaotic Levy flight bat algorithm for parameter estimation in nonlinear dynamic biological systems", *Journal of Computer and Information Technology*, Vol. 2(2), pp. 56-63, February 2015.

[9] Aliasghar Baziar, Abdollah Kavoosi-Fard, and Jafar Zare, "A novel self adaptive modification approach based on bat algorithm for optimal management of renewable MG", *Journal of Intelligent Learning Systems and Applications*, Vol. 5, pp. 11-18, 2013.

[10] Selim Yilmaz and Ecir U. Kucuksille, "Improved bat algorithm (IBA) on continuous optimization problems", lecture notes on Software Engineering, Vol. 1(3), pp. 279-283, August 2013.

[11] A. Kaveh and P. Zakian, "Enhanced bat algorithm for optimal design of skeletal structures", *Asian Journal of Civil Eng.* (BHRC), Vol. 15(2), pp. 179-212, 2014.

[12] Zhen Chen, Yongquan ZhouU, and Mindi Lu, "A simplified adaptive bat algorithm based on frequency", *Journal of Computational Information Systems*, Vol. 9(16), pp. 6451-6458, 2013.

[13] Iztok Fister Jr., Simon Fong, Janez Brest, and Iztok Fister, "A novel hybrid self-adaptive bat algorithm", *Scientific World Journal*, Vol. 2014, Article ID 709738, 12 pages, 2014, http://dx.doi.org/10.1155/2014/709738.

[14] Iztok Fister Jr., Dusan Fister, and Xin-She Yang, "A hybrid bat algorithm", *Elektrotehni Ski Vestnik*, Vol 80(1-2), pp. 1-7, 2013.

[15] Tien-Szu Pan, Thi-Kien Dao, Trong-The Nguyen, and Shu-Chuan Chu, "Hybrid particle swarm optimization with bat algorithm", *Advances in Intelligent Systems and Computing*, Vol. 329, pp. 37–47, 2015.

[16] Yanming Duan, "A hybrid optimization algorithm based on bat and cuckoo search", *Advanced Materials Research*, Vols. 926-930, pp. 2889-2892, 2014.

[17] Xing-shi He, Wen-Jing Ding, and Xin-She Yang, "Bat algorithm based on simulated annealing and Gaussian perturbations", *Neural Computing and Applications*, Vol. 25(2), pp. 459-468, August 2014.

[18] Jeng-Shyang Pan, Vaclav Snasel, Emilio S. Corchado, Ajith Abraham, and Shyue-Liang Wang, "Intelligent Data Analysis and Its Applications, Volume II", Proceeding of the First Euro-China Conference on Intelligent, Data Analysis and Applications, June 13-15, 2014, Shenzhen, China, 2014.

[19] GaigeWang and Lihong Guo, "A novel hybrid bat algorithm with harmony search for global numerical optimization", *Journal of Applied Mathematics*, Vol. 2013, Hindawi Publishing Corporation, Article ID 696491, 21 pages, http://dx.doi.org/10.1155/2013/696491.

[20] Ahmed Majid Taha, and Alicia Y.C. Tang, "Bat algorithm for rough attribute reduction", *Journal of Theoretical and Applied Information Technology*, Vol. 51(1), 2013.

[21] Ahmed Majid Taha, Aida Mustapha, and Soong-Der Chen, "Naive bayes-guided bat algorithm for feature selection", *Scientific World Journal*, Vol. 2013, Article ID 325973, 9 pages, Hindawi Publishing Corporation, http://dx.doi.org/10.1155/2013/325973.

[22] P.R. Srivastava, A. Bidwai, A. Khan, K. Rathore, R. Sharma, and X.S. Yang, "An empirical study of test effort estimation based on bat algorithm", *Int. J. Bio-Inspired Computation*, Vol. 6(1), pp. 57-70, 2014.

[23] Liji Jacob, "Bat algorithm for resource scheduling in cloud computing", *Internal Journal for Research in Applied Science and Engineering Technology*, Vol. 2(IV), pp. 53-57, 2014.

[24] Sonia Goyal and Manjeet Singh Patterh, "Wireless sensor network localization based on bat algorithm", *International Journal of Emerging Technologies in Computational and Applied Sciences* (IJETCAS), Vol. 4(5), pp. 507-512, March-May 2013.

[25] Marwa Sharawi, E. Emary, Imane Aly Saroit, and Hesham El-Mahdy, "Bat swarm algorithm for wireless sensor networks lifetime", *International Journal of Science and Research (IJSR) ISSN* (Online): 2319-7064, Vol. 3(5), pp. 654-664, May 2014.

[26] S. Sakthivel, R. Natarajan, and P. Gurusamy, "Application of bat optimization algorithm for economic load dispatch considering valve point effects", *International Journal of Computer Applications*, Vol. 67(11), April 2013.

[27] Teodoro C. Bora, Leandro dos S. Coelho, and Luiz Lebensztajn, "Bat-inspired optimization approach for the brushless DC wheel motor problem", *IEEE Trans. on Magentics*, Vol. 48(2), pp. 947-950, Feb. 2012.

[28] Ezgi Deniz Ulker and Sadik Ulker, "Microstrip coupler design using bat algorithm", *International Journal of Artificial Intelligence and Applications (IJAIA)*, Vol. 5(1), pp. 127-133, January 2014.

[29] Nazri Mohd. Nawi, Muhammad Zubair Rehman, and Abdullah Khan, "The Effect of Bat Population in Bat-BP Algorithm", The 8th International Conference on Robotic, Vision, Signal Processing and Power Applications, Lecture Notes in Electrical Engineering, Vol. 291, pp. 295-302, 2014.

[30] Sashikala Mishra, Kailash Shaw, and Debahuti Mishra, "A New Meta-Heuristic Bat Inspired Classification Approach for Microarray Data", Procedia Technology, 2nd International Conference on Computer, Communication, Control and Information Technology (C3IT-2012), Vol. 4, pp. 802-806, February 25-26, 2012.

[31] Omar S. Soliman, and Mustafa Abdul Salam, "A hybrid BA-LS-SVM model and financial technical indicators for weekly stock price and trend prediction", International Journal of Advanced Research in Computer Science and Software Engineering, Vol. 4 (4), pp. 764-781, April 2014.

[32] Adis Alihodzic and Milan Tuba, "Bat algorithm (BA) for image thresholding", *Recent Researches in Telecommunications, Informatics, Electronics and Signal Processing*, pp. 364-369, 2014.

[33] V. Rajinikanth, J.P. Aashiha, A. Atchaya, "Gray-level histogram based multilevel threshold selection with bat algorithm", *International Journal of Computer Applications*, Vol. 93(16), May 2014.

[34] Goutam Das, "Bat algorithm based soft computing approach to perceive hairline bone fracture in medical X-ray images", *International Journal of Computer Science and Engineering Technology* (IJCSET), Vol. 4(4), pp.432-436, April 2013.

[35] Guanghui Liu, Heyan Huang, Shumei Wang, and Zhaoxiong Chen, "A novel spatial clustering analysis method using bat algorithm", *International Journal of Advancements in Computing Technology* (IJACT), Vol. 4(20), pp. 561-571, November 2012.

[36] S. Wullf, R. Urner, and S. Ben-David, "Monochromatic bi-clustering", *Journal of Machine Learning Research (JMLR)*, Vol. 28(2), pp. 145-153, 2013.

[37] Adis Alihodzic and Milan Tuba, "Improved bat algorithm applied to multilevel image thresholding", *Scientific World Journal*, Vol. 2014, Article ID 176718, 16 pages, http://dx.doi.org/10.1155/2014/176718, 2014.

[38] Ivanov and Zhdanov, Kaczmarz, "Algorithm for Tikhonov regularization problem", *Applied Mathematics E-Notes*, Vol. 13, pp. 270-276, 2013.

[39] M. M. George, M. Karnan, and R. Siva Kumar, "Supervised artificial bee colony system for tumor segmentation in CT/MRI images", *International Journal of Computer Science and Management Research*, Vol. 2(5), pp. 2529-2533, May 2013.

3 Artificial Fish Swarm

3.1 Fish swarm optimization

3.1.1 Biological base

Artificial fish swarm algorithm (AFSA) is a population-based evolutionary computing (EC) technique that is inspired by the social behavior of fish schooling and swarm intelligence (SI). AFSA, with its artificial fish (AF) concept, was proposed by Dr. Li Xiao-Lei in 2002 [1]. AF adopts information by sense organs and performs stimulant reaction by controlling the tail and fin. The AFSA is a robust stochastic technique in solving optimization problems based on the movement and intelligence of swarms in the food-finding process. Swarms have much explorative capability and noncentralized decision making. The three main principles developed in AFSA are the fish behaviors in food searching, swarming, and following.

3.1.2 Artificial fish swarm algorithm

The main states of a particular fish are *Random*, *Search*, *Swarm*, *Chase*, and *Leap*; see Algorithm 2. The fish swap among one or more states in the same iteration. Details concerning these procedures are presented below.

The main issue of the artificial fish swarm algorithm is the visual scope of each fish. The visual scope of each point X_i is defined as the closed neighborhood of X_i with ray equal to a positive quantity called *visual*. When the visual scope of a fish is empty the fish applies *random* search state. When the fish is in the *random* state it jumps randomly to a new position if the new position has better fitness.

When the visual scope is crowded, the point has some difficulty in following any particular point, and searches for a better region choosing randomly another point (from its visual scope) and moves toward it; *search* behavior. In this case, the fish moves along a direction defined as in the following equation:

$$d_i = X_{rand} - X_i \tag{3.1}$$

X_{rand} is the position of a random fish in the visual scope of fish i.

When the visual scope is neither crowded nor empty, the fish is able either to *swarm* or move toward the central; *swarming* state, or to *chase* or move toward the best point; *chasing* behavior.

input : n Number of Fish in the swarm
          $NIter$ Number of iterations for optimization
          $Visual$ The visual scope radius
          T Leaping Threshold
output: Optimal fish position and its fitness

Initialize a population of n fish' positions at random
;
while *Stopping criteria not met* **do**
    foreach $Fish_i$ **do**
    Compute the visual scope
    if *Visual Scope is empty* **then**
        | $Y^i \leftarrow random(X^i)$
    else
        if *Visual Scope is crowd* **then**
            | $Y^i \leftarrow search(X^i)$
        else
            Calculate Swarm centroid C_i **if** *Fitness of C_i better than Fitness*
            of $Fish_i$ **then**
                | $Y_1^i \leftarrow swarm(X^i)$
            else
                | $Y_1^i \leftarrow search(X^i)$
            end
            if *Fitness of $Fish_i$ worse than global best* **then**
                | $Y_2^i \leftarrow chase(X^i)$
            else
                | $Y_2^i \leftarrow search(X^i)$
            end
            $Y^i = argmin fitness(Y_1^i, Y_2^i)$
        end
    end
    foreach $Fish_i$ **do**
    $X^i \leftarrow select(X^i, Y^i)$
    if *No Advance In global best for T Iterations* **then**
        | Randomly choose a fish X^i $Y^i \leftarrow leap(X^i)$
    end
end

Algorithm 2: Pseudo-code of the AFSA algorithm

Input: x^i, l, u, d^i

1. $\lambda \sim U[0, 1]$

2. For $k = 1, ..., n$ do

   a If $d_k^i > 0$ then

$$y_k^i \leftarrow x_k^i + \lambda \frac{d_k^i}{\|d^i\|}(u_k - x_k^i)$$

   b Else

$$y_k^i \leftarrow x_k^i + \lambda \frac{d_k^i}{\|d^i\|}(x_k^i - l_k)$$

   c End If

3. End For

Algorithm 3: Movement along a particular direction

In the swarming state the fish moves along a direction defined as

$$d_i = C_i - X_i \tag{3.2}$$

where X_c is calculated as $C_i = \frac{\sum_{j \epsilon visual_i} X_j}{np_i^{visual}}$, np_i^{visual} is the number of fish in the visual scope of fish i, and X_j is the position of fish j inside the visual domain of fish i. In the chasing state, the fish moves along a direction defined using the following form:

$$d_i = X_{best} - X_i \tag{3.3}$$

where X_{best} is the best solution found ever. A fish applies a move along a particular direction d_i as shown in Algorithm 3:

A *leaping* behavior can be applied if the best solution gets no advance for a predetermined number of successive iterations T. The leaping behavior randomly selects a point from the population and carries out a random move inside its surround. Algorithm 4 describes the pseudo-code of this leap procedure [2, 5].

3.2 AFSA variants

3.2.1 Simplified binary artificial fish swarm algorithm

A simplified binary version of the AFSA is presented in [3]. In the simplified version, a threshold $0 < \tau1 < \tau2 < 1$ is employed in order to perform the movements of random, searching, and chasing. If $rand < \tau1$, the random behavior is applied. In this behavior the trial point v^l is created by randomly setting 0/1 bits of length m. The chasing behavior is implemented when rand$(0, 1) \geq \tau2$ and it is related to the movement toward the best point found so far in the population, y^{best}. Here, the trial point v^l is created

Input: x, l, u

1. $rand \sim U[0, ..., 1]$

2. For $k = 1, ..., n$ do

   a $\lambda_1 \sim U[0, 1]$; $\lambda_2 \sim U[0, 1]$

   b If $\lambda_1 > 0.5$ then
   $$y_k = x_k^{rand} + \lambda_2(u_k - x_k^{rand})$$

   c Else
   $$y_k = x_k^{rand} - \lambda_2(x_k^{rand} - l_k)$$

   d End If

3. End For

Algorithm 4: Leaping behavior

using a uniform crossover between y^l and y^{best}.

In uniform crossover, each bit of the trial point is created by copying the corresponding bit from one or the other current point with equal probability. The searching behavior is related to the movement toward a point y^{rand} where $rand$ is an index randomly chosen from the set of fish $1, 2, ..., N$. This behavior is implemented when $\tau 1 < rand(0, 1) < \tau 2$. A uniform crossover between y^l and y^{rand} is performed to create the trial point v^l.

A mutation is performed in the point y^{best} to create the corresponding trial point v. In mutation, a 4-flip bit operation is performed, i.e., four positions are randomly selected and the bits of the corresponding positions are changed from 0 to 1 or vice versa. Local search based on swap move is applied after the selection procedure. In this local search, the swap move changes the value of a 0 bit of a current point to 1 and simultaneously another 1 bit to 0. The points in a population may converge to a non-optimal point. To diversify the search, re-initialization is applied to the population randomly, every R iteration keeping the best solution found so far.

3.2.2 Fast artificial fish swarm algorithm (FAFSA)

A fast version of the AFSA has been presented in [6]. In that implementation, the random numbers used in the standard AFSA that are drawn from uniform distribution are replaced by Brownian motion and Lèvy flight, which enhances the convergence speed of the algorithm as pretended by the authors. The Brownian motion is closely linked to the normal distribution. The process has independent increments with expected value 0.

3.2.3 Modified artificial fish swarm optimization

Since most of the behaviors of the fish are local behavior, the algorithm may stuck in local minima [7]. So in [7] a random leaping is presented. A leaping is performed if for a predetermined set of iterations the enhancement in the best fitness is limited; then a random leaping is performed as in equation 3.4.

$$X_i^{t+1} = X_i^t + \beta visual.rand \tag{3.4}$$

where i is the index for a random fish drawn from the swarm, β is a constant, $rand$ is a random number drawn from uniform distribution in the range $[0, 1]$, and $visual$ is the fish visual.

The fish step is also an important parameter in the performance of the AFSA. When step size is small the algorithm can quickly converge to a good solution but can oscillate around the minima. When step size is small the algorithm takes longer time in convergence but convergence to global solution is much ensured. Using the adaptive step may prevent the emergence of vibration, increase the convergence speed, and enhance the optimization precision [7]. In the behaviors of AF-Prey, AF-Swarm, and AF-Follow, which use the Step parameter in every iteration, the optimized variables (vector) have the various quantity of $Step * Rand()$, $Step$ is a fixed parameter, and $Rand()$ is a uniformly distributed function. Zhu and Jiang [11] give an adaptive Step process using the following equation, where t means iteration time:

$$step_{i+1} = \frac{\alpha.N - t}{N} step_i \tag{3.5}$$

where α is a constant between 1.1 and 1.5, and t, N are the iteration number and number of iterations. In the above equation the step size decreases throughout iterations to allow for coarse exploration in the beginning of the optimization and detailed exploration in the end of optimization.

In [28] adaptive $visual$ and $step$ parameters are proposed. In order to control values of step and visual and balancing between global search and local search, a parameter called movement weight (MW) is presented. MW can be a constant value smaller than one, positive linear or nonlinear function. In each iteration, visual and step values are given according to the following equations:

$$visual_t = MW.visual_{t-1} \tag{3.6}$$

$$step_t = MW.step_{t-1} \tag{3.7}$$

where $iter$ is the current iteration of the algorithm. With the purpose of attaining better values for the visual and step based on the iteration number, different methods for calculating MW have been presented in [28].

- Linear movement of MW: In this method, MW is a positive function that varies between a minimum and a maximum and is calculated according to current iteration and final iteration number; see equation 3.8.

$$MW_t = MW_{min} + \frac{Iter_{max} - t}{Iter_{max}}(MW_{max} - MW_{min}) \tag{3.8}$$

where MW_t is the movement at time t, MW_{min}, MW_{max} are the minimum and maximum values for the movement, and $Iter_{max}$ is the total number of iterations in the optimization.

- Random movement of MW: MW is a random number between two values: minimum and maximum. In each iteration, random value of MW is calculated based on the following equation:

$$MW_t = MW_{min} + rand(MW_{max} - MW_{min}) \tag{3.9}$$

where *rand* is a random number drawn from normal distribution between 0 and 1.

Artificial fish swarm algorithm has global search capability, but the convergence speed of later stages is too slow; a lot of artificial fish perform invalid searches which wastes much time [8]. An improved algorithm based on swallowed behavior is proposed in [8]. A diversity indicator is calculated as in equation 3.10 to switch to these two behaviors.

$$\alpha_i = \frac{\min(f, f_{avg})}{\max(f, f_{avg})} \tag{3.10}$$

where f_{avg} is the average fitness of the whole swarm and f is the fitness of the current fish. After a certain time of iteration a threshold is applied to individual fish to decide if it will be swallowed or not. These proposed modifications improve the AFSO algorithm's stability and the ability to search the global optimum. The improved algorithm has more powerful global exploration ability and faster convergence speed, and can be widely used in other optimization tasks especially for high dimensional data.

3.2.4 New artificial swarm algorithm (NAFSA)

In the paper in [9], the following problems are outlined in the standard AFSA:

- *Lack of using previous experiences of AFs during optimization process:* Exploitation performance of AFs decreases. AFs in AFSA do not use the best recorded position on the bulletin for their next movements. This is a considerable weak point in AFSA, since the best result found by swarm so far is not applied for improving swarm member positions.

- *Lack of existing balance between exploration and exploitation:* Setting the initial value for visual and step parameters has a considerable effect on the quality of the final result, because the values of these two parameters remain fixed and are equal to their initial value up to the end of the algorithm execution. By adjusting the initial values of Visual and Step parameters AFs are able to appropriately perform just one of the exploration and exploitation tasks.

- *Wasting high computational load:* There are some complicated calculations such as calculating in two behaviors and applying one only and the cost of calculating the spatial distance between fish.

- *Weak points of AFSAs parameters:* Parameters such as $visual, crowd, step$, are very difficult to select while it affects the performance.

Contraction factor (CF) is added to NAFSA structure. CF is a positive number less than 1 that could be considered constant or be generated by a function. Previously, inertia weight was presented for balancing the exploration and exploitation in PSO. CF in NAFSA has a similar application to inertia weight in PSO. The updating for the CF is defined as follows:

$$CF = CF_{min} + (CF_{max} - CF_{min})rand \tag{3.11}$$

where CF_{max}, CF_{min} are the minimum and maximum allowed value for the CF and $rand$ is a random number drawn from uniform distribution in the range from 0 to 1. The *visual* parameter is calculated given the CF using the following equation:

$$visual_d^{t+1} = visual_d^t * CF \tag{3.12}$$

where CF is as calculated in equation 3.11 and $visual_d^t$ is the *visual* parameter in dimension d at iteration t.

In the NAFSA the new behaviors are individual behavior and group behavior. The *individual behavior* is composed of prey and free move behaviors. In this case, AF_i which is in position $X_i(t)$, tries a number of times to move toward better positions. In each try, AF considers position $X_j(t)$ around itself by equation. 3.13, then evaluates fitness value of that point. If better than current position $X_i(t)$ then it will be kept else a new trial is applied. If for a number of failure trials no better solution is found then a random position is chosen according to equation 3.13.

$$x_{j,d}^{t+1} = x_{i,d}^t + visual.rand_d(-1, 1) \tag{3.13}$$

where $rand_d(-1, 1)$ is a random number drawn from uniform distribution in the range [-1,1], and *visual* is AFSA visual parameter. In *group behavior*, two goals are followed: first goal is to keep AFs as a swarm and the second one is the movement of AFs toward the best AFs position. Central position of swarm is calculated by equation 3.14.

$$x_{c,d} = \frac{1}{N} \sum_{i=1}^{N} x_{i,d} \tag{3.14}$$

where N is the number of fish in the swarm, d is the dimension, and s_c is the central position. Movement condition toward center position makes use of x_c; also, if it gets to a worse solution it tries to move toward the best solution.

3.2.5 AFSA with mutation

An efficient artificial fish swarm (AFS) algorithm should be capable of exploring the whole search space as well as exploiting around the neighborhood of a reference point, for example, the best point of the population [29]. Previous experiments have shown that AFS algorithms may be trapped into local optimum, although the leaping behavior aims at jumping out from local solutions. The mutation operator has been used as a diversifying procedure to translate three important behaviors in AFS algorithm: searching, random, leaping [29]. The introduction of the mutation operator into the AFS algorithm aims at diversifying the search, thus preventing the algorithm from falling into a local optimum. The mutated artificial fish swarm algorithm (MutAFSA for short) uses the following mutation strategies:

$$y_i = \begin{cases} x_i + F_1(x_{best} - x_i + x_{r1} - x_{r2}), & \text{for searching behavior} \\ x_i + F_2(x_{r1} - x_i) + F_1(x_{r2} - x_{r3}), & \text{for random behavior} \\ x_i + F_1(x_{r2} - x_{r3}), & \text{for the leaping behavior} \end{cases} \quad (3.15)$$

where $r_j, j = 1, 2, 3$ are randomly chosen indices from the set $1, ..., p_{size}$, mutually different and different from the running index i. F_1 and F_2 are real and constant parameters from $[0,1]$ and $[0,2]$, respectively.

3.2.6 Fuzzy adaptive AFSA

In nature, fish swarm members have a certain visual that directly depends on the fish type, environment conditions (e.g., water fog), and nearn obstacles (e.g., water plants and other fish). When the swarm moves toward a target (e.g., food) as much of it converges, visibility is reduced due to density [4]. In order to control values of step, visual, and balancing between global search and local search, a novel parameter, called *Constriction Weight*, is proposed. Weight has to be greater than 0 and smaller than 1. Current iteration visual and step values are calculated according to the following formulas in the presence of weight parameters:

$$visual_{iter} = CW\, visual_{iter-1} \quad (3.16)$$

$$step_{iter} = CW\, step_{iter-1} \quad (3.17)$$

With the purpose of attaining better values for the visual and step two different fuzzy methods for calculating the weight have been proposed in [4].

1. In fuzzy uniform fish CW weight is a value between 0 and 1 that is calculated as an output of the fuzzy engine. All of the fish in the swarm then adjust their

visual and step based on the output weight. The proposed fuzzy engine has two inputs and one output: *iteration number* and *ratio of improved fish* as inputs and constriction weight as an output.

Iteration number, normalized between 0 and 1, is the proportion of the current iteration number to the final iteration number. *Ratio of improved fish* is the proportion of the number of fish that finds better positions in problem space (points with higher fitness) to the total number of fish in comparison with a previous iteration.

Based on the above two inputs the CW is calculated by a fuzzy system defined using the Mamdani fuzzy system.

2. In uniform autonomous (FUA) each artificial fish adjust its visual and step parameters individually and independent of the rest rest of the swarm. The input to the fuzzy system in this case are distance from best, fitness ranking and iteration number. Distance from best is a normalized rank-based value between 0 and 1 for each fish. Fitness ranking is equal to the proportion of the ranking number, calculated based on the fitness value for the artificial fish, to the total number of artificial fish. Also, the Mamdani fuzzy system is constructed based on these variables to output the value of CW.

3.2.7 AFSA with adaptive parameters

There are many parameters that need to be adjusted in AFSA [10]. Among these parameters, visual and step are very significant in view of the fact that artificial fish basically move based on these parameters. Always, these two parameters remain constant until the algorithm termination. Large values of these parameters increase the capability of algorithm in a global search, while small values improve the local search ability of the algorithm. The work in [10] empirically studies the performance of the AFSA, and different approaches to balance between local and global exploration have been tested based on the adaptive modification of visual and step during algorithm execution.

In order to control values of step and visual and balancing between global search and local search, authors of [10] proposed a parameter called Movement Weight (MW). MW can be a constant value smaller than one, or a positive linear or nonlinear function. In each iteration, visual and step values are given according to the following equations:

$$Visual_i = MW \times Visual_{i-1} \tag{3.18}$$

$$Step_i = MW \times Step_{i-1} \tag{3.19}$$

where i is the current iteration of the algorithm. The MW parameter can be updated using the following linear formula:

$$MW_i = MW_{min} + \frac{Iter_{max} - i}{MW_{max} - MW_{min}} \times (MW_{max} - MW_{min}) \tag{3.20}$$

where MW_{min}, MW_{max} are the allowed minimum and maximum value for MW and $Iter_{max}, i$ are the maximum allowed number of iteration and the current iteration number in order.

A random updating mechanism was also proposed in [10] as follows:

$$MW_i = MW_{min} + rand \times (MW_{max} - MW_{min}) \tag{3.21}$$

where MW_{min}, MW_{max} are the allowed minimum and maximum value for MW and *rand* is a random number.

3.2.8 AFSA with modified preying

Preying behavior is the basic behavior of AFSA. To further enhance the performance of AFSA, the preying behavior is modified as follows [23]: Let X_i be the artificial fish current state and select a state X_j randomly in its visual distance and let Y be the food density (objective function value). If $Y_i > Y_j$, the fish goes forward a step in the direction of the vector sum of the X_j and the best X_{best}. X_{best} is the best artificial fish state till now. Otherwise, select a state X_j randomly again. If it cannot satisfy after N times, it moves a step randomly.

3.2.9 Quantum AFSA

A modified quantum-based AFSA was proposed in [11]. The goal of the hybridization is to improve the global search ability and the convergence speed of the artificial fish swarm algorithm (AFSA). The hybrid algorithm is based on the concepts and principles of quantum computing, such as the quantum bit and quantum gate. The position of the artificial fish (AF) is encoded by the angle in $[0, 2\pi]$ based on the qubit's polar coordinate representation in the two-dimension Hilbert space. The quantum rotation gate is used to update the position of the AF in order to enable the AF to move and the quantum non-gate is employed to realize the mutation of the AF for the purpose of speeding up the convergence.

3.2.10 Chaotic AFSA

In [26] a chaotic version of AFSA was presented. Chaotic phenomenon is a specific phenomenon of the nonlinear dynamic system with randomness, ergodicity, deterministic properties. Because the chaotic searching can be easily executed and can avoid the local extremes and it is better than random searching, the chaotic variables have advantages for local searching. Chaotic sequence is generated for every fish in every dimension by projecting and normalizing the fish's position on that dimension and then the chaotic sequence is generated. The generated chaotic sequence is projected back to the original space so that it can be evaluated. A selection phase is used to compare the fish's original and chaotic position and the better position is selected as a candidate position for the AFSA.

The chaotic variables use the tent chaotic map defined as follows:

$$x_{i+1} = \begin{cases} 2x_i & x_i \in (0, 0.5] \\ 2(1 - x_i) & x_i \in (0.5, 1] \end{cases} \tag{3.22}$$

3.2.11 Parallel AFSA (PAFSA)

A parallel version of AFSA was introduced in [12]. Because the main loop of the basic AFSA only selects one of the three behaviors (chasing behavior, the swarming behavior, and the feeding behavior) to execute the result is easy to be stagnant or be missed, and it is not very satisfactory. Therefore, we propose an improved parallel artificial fish swarm algorithm (PAFSA) which reduces the probability of missing the better solutions. The improvement of PAFSA is made at the loop body. This method is divided into two paths after initialization execution: one path to perform the chasing behavior (feeding behavior is a random behavior), and another path to perform the swarming behavior (and the feeding behavior is also a random behavior), by comparing the fitness value of the two behaviors, choosing the better result and recording on the bulletin board at the same time to update the individual to continue the iterative optimization.

3.3 AFSA hybridizations

3.3.1 AFSA hybrid with culture algorithm (CA) (CAFAC)

A hybrid algorithm is proposed in [14] for enhanced optimization. The AFSA is tolerant to initial value and parameter settings. However, due to the random step length and random behavior, the artificial fish converges slowly at the mature stage of the algorithm. Thus, the optimization precision usually cannot be high enough, which results in the blindness of search and poor ability of maintaining the balance of exploration and exploitation. Crossover is added to allow the fish to jump to and explore a diverse area. The exploitation of the crossover operator can help the artificial fish to not only jump out the blindness search but also to inherit the advantage of its parents. Inspired by the principle of the CA, they applied the normative knowledge and situational knowledge stored in belief space in the CA into the AFA. The fish swarm is regarded as the population space, from where the domain knowledge is extracted. Hence, the domain knowledge is formed and stored in belief space so as to model and impact the evolution of the population at iteration.

The following equation is used to decide if the algorithm is stuck in local minima:

$$\|\frac{f(x_i^t - f(x_i^{t-1})}{f(x_i^{t-1})}\| < 0.1 \tag{3.23}$$

when the above criterion is satisfied crossover is applied on the *ith* artificial fish as in equation 3.24.

$$x_i^{new} = x_{r1} + \alpha(x_{r2} - x_{r1}) \tag{3.24}$$

where x_{r1}, x_{r2} are two random fish and $r1 \neq r2 \neq i$. α is a random number uniformly distributed in interval of $[1-d, 1+d]$, and d is chosen as 0.25. Evaluate the child x_i, and replace the individual x_i^{new} with the child if its fitness is better. The main steps for the CAFAC are:

1. Set all the values for the parameters, and initialize the N artificial fish in the search ranges with random positions.

2. Evaluate all the artificial fishes using the fitness function y, and initialize the belief space.

3. For each *ith* artificial fish, simulate the preying pattern, swarming, and chasing patterns separately, and select the best child fish. If the child is better, replace the *ith* artificial fish with the child.

4. Update the belief space.

5. If the crossover criterion is satisfied, apply the crossover operator to the *ith* artificial fish from Step 3.

6. Return to Step 3 until the termination criterion is satisfied.

3.3.2 AFSA hybrid with PSO

A hybridization between PSO and AFSA based on the number of fellows was proposed in [15]. af parameter was defined as the number of fellows of a given fish in its visual scope. When this number is high, this means that there is enough information around the fish to continue exploration using AFSA. When the af is small the standard PSO is applied. PSO can search the best position by the best position of particles and swarm on the call-board.

3.3.3 AFSA hybrid with glowworm optimization

In [16], a hybrid algorithm was proposed between AFSA and the glow worm optimizer. In the basic GSO algorithm, each glowworm—only in accordance with luciferin values of glowworms in its neighbor set—selects the glowworm by a certain probability and moves toward it. If the search space of a problem is very large or irregular, the neighbor sets of some glowworms may be empty, which leads these glowworms to keep still in iterative process. To avoid this case and ensure that each glowworm keeps moving, the predatory behavior of AFSA has been hired into glowworm swarm optimization (GSO). The idea of hybrid algorithm is that the glowworms whose neighbor sets are empty carries out predatory behavior in their dynamic decision domains.

3.3.4 AFSA hybrid with cellular learning automata

A hybrid version of AFSA with cellular automata was proposed in [17]. In the proposed algorithm, each dimension of search space is assigned to one cell of cellular learning automata and in each cell a swarm of artificial fish are located which have the optimization duty of that specific dimension. In fact, in the proposed algorithm for optimizing D-dimensional space, there are D one-dimensional swarms of artificial fishes and each swarm is located in one cell and they contribute with each other to optimize the D-dimensional search space. The learning automata in each cell is responsible for making diversity in artificial fishes swarm of that dimension and equivalence between global search and local search processes.

3.4 Fish swarm in real world applications

This section presents sample applications for artificial fish swarm algorithm. AFSA was successfully applied in many disciplines in decision support and making, image processing domain, engineering, etc. The main focus is on identifying the fitness function and the optimization variables used in individual applications. Table 3.1 summarizes sample applications and their corresponding objective(s).

Grid computing is a high performance computing environment to solve larger scale computational demands. Grid computing contains resource management, job scheduling, security problems, and information management. Job scheduling is a fundamental issue in achieving high performance in grid computing systems. It is a big challenge for efficient scheduling algorithm design and implementation [7]. Unlike scheduling problems in conventional distributed systems, this problem is much more complex as new features of grid systems such as its dynamic nature and the high degree of heterogeneity of jobs and resources must be tackled. The problem is multi-objective in its general formulation, the two most important objectives being the minimization of make-span and flow-time of the system. Job scheduling is known to be NP-complete [7] which motivates using AFSA to optimize it. A modified version of AFSA was used in [7] for the efficient job scheduling.

AFSA is used in [18] to train a feed-forward neural network. Each artificial fish represents a feed-forward neural network. The optimizing variables are weight matrix and biases. The feed-forward neural network (NN) training process is to get the minimum value of network error, E, by adjusting the weights and biases values. The nonlinear error function chosen is mean square error to quantify the error of the network.

In [24] AFSA is used as both a feature selection method and neural trainer at the same time. The neural weights and the selected feature set are the target of the optimization. The fitness function is calculated as the function achieving minimum error on the given weight set and feature set.

Table 3.1: AFSA applications and the corresponding fitness function

Application Name	Fitness Function	Sample References
Job scheduling	The minimization of make-span and flow-time of the system	[7]
Training of feed-forward neural network	Fitness function minimize, the neural network output error	[18, 24]
Feature selection	Feature combination with maximum classification performance	[24]
Radial basis function neural network	Minimize over all network error	[13]
Support vector machine training (SVM)	Fitness is performance of SVM	[12]
Feature selection	Rough-set classification performance	[25]
Intrusion detection	Increase the detection rate	[26]
Decision support (taxi allocation to city locations)	The purpose is to obtain an optimal match of the empty taxis and passengers predicted according to the historical scheduling data	[19]
Un-capacitated facility location problem (UFLP)	Minimize the cost of traveling vehicles	[3]
Data clustering	Minimize intracluster distance inside individual clusters	[20]
Motion estimation	Minimum spectral distance between matched blocks	[23]
Welded beam design problem	Minimum cost subject to constraints on shear stress, bending stress in the beam, buckling load on the bar, end deflection of the beam, and side constraints	[21]
Optimization operation of cascade reservoirs	Maximal gross power generation in one scheduling period as the target	[21]
Tension/compression string design problem	Minimization of the weight of a tension/compression spring subject to constraints on minimum deflection, shear stress, surge frequency, limits on outside diameter and on side constraints	[21]
The pressure vessel design problem	Minimization of the total cost of the specimen studied, including the cost of the material, forming, and welding	[21]
Integral plus derivative (PID) controller	System performance maximization	[27]

AFSA was used in combination with radial-basis neural network (RBFF) to predict the stock indices in [13]. The movement of stock index is difficult to predict for it is non-linear and subject to many inside and outside factors. That work selects radial basis functions neural network (RBFNN) to train data and forecast the stock index in the Shanghai Stock Exchange. In order to solve the problem of slow convergence and low accuracy, and to ensure better forecasting result, they introduce artificial fish swarm algorithm (AFSA) to optimize RBF, mainly in parameter selection. Empirical tests indicate that RBF neural network optimized by AFSA can have ideal results in short-term forecast of stock indices.

In [12] AFSA was employed in the training of the support vector machine (SVM). A parallel version of AFSA was employed to find the optimal parameter for SVM, namely, kernel and penalty factors. The appropriate kernel functions and parameters are selected, according to the specific circumstances of the problem. Gaussian kernel has better adaptability; whether it is low-dimensional, high-dimensional, smaller sample, or larger sample, Gaussian kernel function is applicable and it is a satisfactory function for classification. In that work, the Gaussian kernel function is selected as the kernel function of SVM. Thus, the target of optimization is to select the Gaussian kernel parameter and the penalization factor given the SVM performance as a fitness function.

A system for feature selection based on AFSA was proposed in [25]. The proposed algorithm uses rough set to assess the quality of individual fish, while it uses the searching capability of AFSA to search for optimal feature combination. The proposed algorithm searches the minimal reduct in an efficient way to observe the change of the significance of feature subsets and the number of selected features.

The anomaly detection or abnormal intrusion detection can detect the new attack modes which have become the main direction of current research [26]. The network intrusion detection is a pattern recognition problem in which the original network intrusion features contain redundant features that cannot only cause a bad effect on the performance of the classifier and increase the probabilities of dimension disaster appearance influencing the efficiency, but also make the detection accuracy more worse when the amount of the features excess some threshold because there are no linear proportional relationships between the amount of intrusion features and detection results. Thus, it is important to select the key features strongly related to the intrusion features and eliminate the redundancy. The objective functions are constructed according to the feature subset dimensions and the detection accurate rates of the detection model. Then the artificial fish swarm algorithm is used to search the optimal feature subset.

The purpose of optimizing the dispatch is to guide the empty taxis to arrive the position where passengers may be waiting for the taxis [19]. An improved artificial fish swarm algorithm (AFSA)-based method is proposed in [19] for the optimization of a city taxi scheduling system. In this method, the grid scheduling algorithm is used for taxi automatic scheduling, and the AFSA is used to further optimize this scheduling system.

The purpose is to obtain an optimal match of the empty taxis and passengers predicted according to the historical scheduling data.

The un-capacitated facility location problem (UFLP) is faced in [3] using AFSA. The UFLP involves a set of customers with known demands and a set of alternative candidate facility locations. If a candidate location is to be selected for opening a facility, a known fixed setup cost will be incurred. Moreover, there is also a fixed known delivery cost from each candidate facility location to each customer. The goal of UFLP is to connect each customer to exactly one opened facility in the way that the sum of all associated costs (setup and delivery) is minimized. It is assumed that the facilities have sufficient capacities to meet all customer demands connected to them. Given the UFLP, the number of alternative candidate facility locations is m and the number of customers is n. The mathematical formulation of the UFLP is given as follows:

$$\downarrow z(x, y) = \sum_{i=1}^{m} \sum_{j=1}^{n} c_{ij} x_{ij} + \sum_{i=1}^{m} f_i y_i \tag{3.25}$$

subject to

$$\sum_{i=1}^{m} x_{ij} = 1 \text{for all } j \tag{3.26}$$

$$x_{ij} \leq y_i \text{for all } i, j \tag{3.27}$$

$$x_{i,j}, y_i \in 0, 1 \text{for all } i, j \tag{3.28}$$

where c_{ij} = the delivery cost of meeting customer $j's$ demand from a facility at location i

f_i = the setup cost of facility at location i.

$$x_{i,j} = \begin{cases} 1, & \text{if customer j is served from location i} \\ 0, & \text{otherwise} \end{cases} \tag{3.29}$$

$$y_i = \begin{cases} 1, & \text{if a facility is opened at location i} \\ 0, & \text{otherwise} \end{cases} \tag{3.30}$$

$$\tag{3.31}$$

In [20] AFSA was used for data clustering, where the intra-cluster variation is used as a fitness and cluster centers are the dimensions to be searched for a hybrid clustering algorithm proposed based on AFSA and k-means approaches. Several advantages including particularly high computational speed and efficient local search are involved in K-means. However, its convergence is extremely sensitive to the chosen initial solution. The goal of the hybrid approach is to improve convergence rate, by conquering the mentioned problems. In that proposed approach, k-means has been applied as a behavior for artificial fish. After executing group and individual behaviors on AFSA, they perform an iteration of k-means on a specified percentage of artificial fish to enhance its local searching.

Motion estimation plays a key role in H.264/AVC video coding, but it is the most time-consuming task in the encoding process. In order to reduce the computational complexity of motion estimation in H.264/AVC video coding, the work in [23] proposes a new search algorithm based on artificial fish swarm algorithm (AFSA) which is a new efficient optimizing method. A modified fast version of AFSA was employed to find the optimal frame shifting in x, y directions. The objective function calculates the residual error for the given block at the given shift.

The welded beam design problem is the process in which a welded beam is designed for minimum cost subject to constraints on shear stress, bending stress in the beam, buckling load on the bar, end deflection of the beam, and side constraints. This design problem was tackled by AFSA in [21].

The optimization operation of cascade reservoirs takes the power discharges as the decision variables, and takes the maximal gross power generation in one scheduling period as the target. The objective function for cascade reservoirs is

$$maximize \text{ E } = 3600. \sum_{p=1}^{n} \sum_{t=1}^{T} (Q_p t . M_t / R_{pt}) \tag{3.32}$$

where E is the gross power generation of cascade power stations, Q_{pt} is the power discharge of power station p in interval t, M_t is the number of hours in interval t, R_{pt} is the mean water rate of power station p in interval t, its unit is m^3/kWh; n is the number of the reservoirs, T is total time-interval for calculation. This problem is constrained to

- Water balance equations constraint

- Storage capacity constraint

- Correlation equation

- Turbine discharge constraint

- Spillway capacity constraint

- Boundary constraint

- Outflow from reservoir constraint

- Output of power station constraint

- Relationship between water level and storage capacity

Tension/compression string design problem was tackled by AFSA in [21]. It is devoted to the minimization of the weight of a tension/compression spring subject to constraints on minimum deflection, shear stress, surge frequency, and limits on outside diameter and on side constraints. The design variables are the wire diameter, the mean coil diameter, and the number of active coils.

The pressure vessel design problem [21] is devoted to the minimization of the total cost of the specimen studied, including the cost of the material, forming, and welding. A cylindrical vessel is capped at both ends by hemispherical heads. There are four design variables: thickness of the shell, thickness of the head, inner radius, and length of the cylindrical section of the vessel, not including the head. This problem was solved by AFSA in [21] and proved good performance.

A three-term Proportional plus Integral plus Derivative (PID) controller, playing a vital role in the field of automation and control, has been counted as a reliable component of industry because of its simplicity and satisfactory performance with vast range of processes. Because of its simple structure, it can be easily understood and implemented in practice. Thus their presence is highly appreciated in practical applications. AFSA was employed in [27] for PID controller design where the parameters K_p, K_i, K_d are to be found by the AFSA. The performance of system is the fitness function for the PID design.

3.4.1 Fish swarm in selection of optimal cluster heads (CHs) locations in wireless network

The battery supplies power to the complete sensor node. Sensor nodes are generally small, light, and cheap. The size of the battery is limited and it plays a vital role in determining sensor node lifetime. The amount of power drawn from a battery should be carefully monitored. AA batteries normally store 2.2 to 2.5 Ah at 1.5 V. However, these numbers vary depending on the technology utilized. For example, Zincair-based batteries have higher capacity in Joules/cm^3 than lithium batteries. Alkaline batteries have the smallest capacity, normally around 1200 J/cm^3. Furthermore, sensors must have a lifetime of months to years, since battery replacement is not an option for networks with thousands of physically embedded nodes. This causes energy consumption to be the most important factor in determining sensor node lifetime. One of the main design goals of WSN is to carry out data communication while trying to prolong the lifetime of the network and prevent connectivity degradation by employing aggressive energy management techniques.

In this case study, an optimized hierarchical routing technique which aims to reduce the energy consumption and prolong network lifetime is presented. In this study, the selection of optimal cluster heads (CHs) locations is based on artificial fish swarm algorithm (AFSA). Various behaviors in AFSA such as preying, swarming, and following are applied to select the best locations of CHs. A fitness function is used to compare between these behaviors to select the best CHs [30].

Each AFSA behavior outputs a set of CHs. To choose the best CHs set, the fitness function is used. The fitness function is described in equation 3.33. The smallest fitness represents the best CHs set among others.

$$fitness = \alpha f_1 + (1 - \alpha) f_2 \tag{3.33}$$

$$f_1 = \frac{\sum_{i=1}^{M} E(n_i)}{\sum_{i=1}^{M} E(CH_m)} \tag{3.34}$$

$$f_2 = \max \sum_{i \in m} \frac{d(n_i, CH_m)}{|c_m|} \tag{3.35}$$

where f_1 is the energy representation part, and it is equal to the sum of all member node energy $E_{(n_i)}$ (not including CH) divided by the sum of all CH energy $E(CH_m)$ referring to equation 3.34. Where f_2 represents the density it is equal to cluster with highest average distance between CH and joined member nodes $d(n_i, CH_m)$ divided by the total member nodes in the same cluster $\|Cm\|$ as shown in equation 3.35.

3.4.2 Fish swarm in community detection in social networks

A social network is a graph made of nodes that are connected by one or more specific types of relationships, such as values, friendship, work. The goal of community detection in networks is to identify the communities by only using the information embedded in the network topology. The problem has a long tradition and it has appeared in various forms in several disciplines. Many methods have been developed for the community detection problem. These methods use tools and techniques from disciplines like physics, biology, applied mathematics, and computer and social sciences [31]. One of the special interests in social network analysis is finding community structure. Community is a group of nodes that are tightly connected to each other and loosely connected with other nodes. Community detection is the process of network clustering into similar groups or clusters. Community detection has many applications including realization of the network structure, detecting communities of special interest, and visualization [31].

A social network can be modeled as a graph $G = (V, E)$, where V is a set of nodes, and E is a set of edges that connect two elements of V. A community structure S in a network is a set of groups of nodes having a high density of edges among the nodes and a lower density of edges between different groups. The problem of detecting k communities in a network, where the number k is unknown, can be formulated as finding a partitioning of the nodes in k subsets that best satisfy a given quality measure of communities $\mathbf{F}(S)$. The problem can be viewed as an optimization problem in which one usually wants to optimize the given quality measure $\mathbf{F}(S)$. A single objective optimization problem $(\Omega;\mathbf{F})$ is defined as in equation 3.36.

$$min \ \mathbf{F}(\boldsymbol{S}), \ s.t \ \boldsymbol{S} \in \Omega \qquad (3.36)$$

where $\mathbf{F}(\boldsymbol{S})$ is an objective function that needs to be optimized, and $\Omega = \{\mathbf{S}_1, \mathbf{S}_2 \ldots \mathbf{S}_k\}$ is the set of feasible community structures in a network. We assume that all quality measures need to be minimized without loss of generality.

3.5 Chapter conclusion

The basic concepts of artificial fish swarm algorithm were discussed in this chapter. In addition, this chapter discusses variants and hybridization with other optimization techniques. We showed also how the artificial fish swarm algorithm was applied to solve real life applications for selection of optimal cluster heads (CHs) locations in wireless network and for community detection in social networks.

Bibliography

[1] X.L. Li, Z.J. Shao, and J.X. Qian, "An optimizing method based on autonomous animates: fish-swarm algorithm", *Methods and Practices of System Engineering*, pp. 32-38, 2002.

[2] Edite M. G. P. Fernandes, Tiago F. M. C. Martins, and Ana Maria A. C. Rocha, "Fish Swarm Intelligent Algorithm for Bound Constrained Global Optimization", International Conference on Computational and Mathematical Methods in Science and Engineering (CMMSE), 30 June – 3 July, Gijún (Asturias), Spain, 2009.

[3] Abul Kalam Azad, Ana Maria A.C. Rocha, and Edite M.G.P. Fernandes, "A Simplified Binary Artificial Fish Swarm Algorithm for Incapacitated Facility Location Problems", Proceedings of the World Congress on Engineering, 2013 Vol I, WCE 2013, July 3 – 5, London, 2013.

[4] Danial Yazdani, Adel Nadjaran Toosi, and Mohammad Reza Meybodi, "Fuzzy Adaptive Artificial Fish Swarm Algorithm", Lecture Notes in Computer Science, Springer-Verlag Berlin Heidelberg, Vol. 6464, pp 334-343, 2011.

[5] Nebojsa Baccanin, Milan Tuba, and Nadezda Stanarevic, "Artificial Fish Swarm Algorithm for Unconstrained Optimization Problems", Proceedings of the 6th WSEAS International Conference on Computer Engineering and Applications, and Proceedings of the 2012 American Conference on Applied Mathematics, pp. 405-410, 2012.

[6] M.A. Awad El-bayoumy, M. Z. Rashad, M. A. Elsoud, and M. A. El-dosuky, "FAFSA: fast artificial fish swarm algorithm", *International Journal of Information Science and Intelligent System*, Vol. 2(4), pp. 60-70, 2013.

[7] Saeed Farzi, "Efficient job scheduling in grid computing with modified artificial fish swarm algorithm", *International Journal of Computer Theory and Engineering*, Vol. 1(1), pp. 1793-8201, April 2009.

[8] Zhehuang Huang and Yidong Chen, "An improved artificial fish swarm algorithm based on hybrid behavior selection", *International Journal of Control and Automation*, Vol. 6(5), pp. 103-116, 2013.

[9] D. Yazdani, B. Saman, A. Sepas-Moghaddam, F. Mohammad-Kazemi, and M. Reza Meybodi, "A new algorithm based on improved artificial fish swarm algorithm for data clustering", *International Journal of Artificial Intelligence*, Vol. 11(A13), pp. 193-221, 2013.

[10] Reza Azizi, "Empirical study of artificial fish swarm algorithm", *International Journal of Computing, Communications and Networking*, Vol. 3(1), January 2014.

[11] K. Zhu and M. Jiang, "Quantum artificial fish swarm algorithm". In: IEEE 8th World Congress on Intelligent Control and Automation, July 6-9, Jinan, China, 2010.

[12] Jing Bai, Lihong Yang, and Xueying Zhang, "Parameter optimization and application of support vector machine based on parallel artificial fish swarm algorithm", *Journal of Software*, Vol. 8(3), pp. 673-679, 2013.

[13] Dongxiao Niu, Wei Shen, and Yueshi Sun, "RBF and artificial fish swarm algorithm for short term forecast of stock indices", 2nd International Conference on Communication Systems, Networks and Applications, June 29, 2010-July 1, 2010, Hong Kong, pp. 139-142, 2010.

[14] Ying Wu, Xiao-Zhi Gao, and Kai Zenger, "Knowledge-based artificial fish-swarm algorithm", 18th IFAC World Congress Milano (Italy) August 28 - September 2, pp. 14705-14710, 2011.

[15] C. Huadong, W. Shuzong, L. Jingxi, and L. Yunfan, "A hybrid of artificial fish swarm algorithm and particle swarm optimization for feed forward neural network training", International Conference on Intelligent Systems and Knowledge Engineering (ISKE07), 2007.

[16] D. Yazdani, S. Golyari, and M.R. Meybodi, "A new hybrid algorithm for optimization based on artificial fish swarm algorithm and cellular learning automata". In: IEEE 5th International Symposium on Telecommunications (IST), 4-6 Dec. 2010, Tehran, pp. 932-937, 2010.

[17] D. Yazdani, S. Golyari, and M.R. Meybodi, A new hybrid algorithm for optimization based on artificial fish swarm algorithm and cellular learning automata. In: IEEE 5th International Symposium on Telecommunications (IST), 4-6 Dec. 2010, Tehran, pp. 932-937, 2010.

[18] Shafaatunnur Hasan, Tan Swee Quo, Siti Mariyam Shamsuddin, and Roselina Salle-huddin, "Artificial neural network learning enhancement using artificial fish swarms algorithm", Proceedings of the 3rd International Conference on Computing and Informatics, ICOCI2011, 8-9 June, Bandung, Indonesia, pp. 117-122, 2011.

[19] Xianmin Ma and Ni Liu, "An improved artificial fish-swarm algorithm based optimization method for city taxi scheduling system", *Journal of Computational Information Systems*, Vol. 9(16), pp. 6351-6359, 2013.

[20] D. Yazdani, B. Saman, A. Sepas-Moghaddam, F. Mohammad-Kazemi, and M. Reza Meybodi, "A new algorithm based on improved artificial fish swarm algorithm for data clustering", *International Journal of Artificial Intelligence*, Vol. 11(A13), pp. 193-221, 2013.

[21] Fran Srgio Lobato and Valder Steffen Jr., "Fish swarm optimization algorithm applied to engineering system design", *Latin American Journal of Solids and Structures*, Vol. 11, pp. 143-156, 2014.

[22] Yong Peng, "An improved artificial fish swarm algorithm for optimal operation of cascade reservoirs", *Journal of Computers*, Vol.6(4), pp. 740-746, April 2011.

[23] Chun Fei, Ping Zhang, and Jiamping Li, "Motion estimation based on artificial fish-swarm in H.264/AVC coding", WSEAS Transactions on Signal Processing, Vol. 10, pp. 221-229, 2014.

[24] M. Zhang, C. Shao, F. Li, Y. Gan, and J. Sun, "Evolving neural network classifiers and feature subset using artificial fish swarm". In: Proceedings of the 2006 IEEE International Conference on Mechatronics and Automation, Luoyang, China, June 25-28, pp. 1598-1602, 2006.

[25] Fei Wang, Jiao Xu, and Lian Li, "A novel rough set reduct algorithm to feature selection based on artificial fish swarm algorithm", *Advances in Swarm Intelligence* Lecture Notes in Computer Science, Vol. 8795, pp. 24-33, 2014.

[26] Liang Lei, "The network abnormal intrusion detection based on the improved artificial fish swarm algorithm features selection", *Journal of Convergence Information Technology*, Vol. 8(6), pp. 206-212, 2013.

[27] Wafa Ali Soomro, I. Elamvazuthi, M.K.A. Ahamed Khan, Suresh Muralidharan, and M. Amudha, "PID controller optimization using artificial fish swarm algorithm", *International Journal of Electrical and Electronics Research*, Vol. 1(1), pp. 11-18, 2013.

[28] Reza Azizi, "Empirical study of artificial fish swarm algorithm", *International Journal of Computing, Communications and Networking*, Vol. 3(1), pp. 1-7, 2014.

[29] A.M.A.C. Rocha and E.M.G.P. Fernandes, "On hyperbolic penalty in the mutated artificial fish swarm algorithm in engineering problems". In: 16th Online World Conference on Soft Computing in Industrial Applications (WSC16), pp. 1-11, 2011.

[30] Asmaa Osama Helmy, Shaimaa Ahmed, and Aboul Ella Hassenian, "Artificial fish swarm algorithm for energy-efficient routing technique", *Intelligent Systems'2014, Advances in Intelligent Systems and Computing*, Vol. 322, pp. 509-519, 2015.

[31] Eslam Ali Hassan, Ahmed Ibrahem Hafez, Aboul Ella Hassanien, and Aly A. Fahmy, "Community detection algorithm based on artificial fish swarm optimization", *Intelligent Systems' 2014, Advances in Intelligent Systems and Computing*, Vol. 323, pp. 509-521, 2015.

4 Cuckoo Search Algorithm

4.1 Cuckoo search (CS)

Cuckoo search is a meta-heuristic algorithm proposed by Yang and Deb in 2009 [1, 2] for solving continuous optimization problems. This algorithm is based on the obligate brood parasitic behavior of some cuckoo species.

4.1.1 Cuckoo breeding behavior

Cuckoo birds have aggressive reproduction strategy. Some species such as the ani and Guira cuckoos lay their eggs in communal nests, though they may remove others eggs to increase the hatching probability of their own eggs. A number of species engage the obligate brood parasitism by laying their eggs in the nests of other host birds [1]. Some host birds can engage direct conflict with the intruding cuckoos. If a host bird discovers the eggs are not its own, it will either throw these alien eggs away or simply abandon its nest and build a new nest elsewhere [1]. Some cuckoo species such as the New World brood-parasitic Tapera are often very specialized in the mimicry in color and pattern of the eggs of a few chosen host species. This reduces the probability of their eggs being abandoned and thus increases their reproductivity [1]. Parasitic cuckoos often choose a nest where the host bird just laid its own eggs to increase the cuckoo chicks' share of food provided by its host bird [1].

4.1.2 Artificial cuckoo search

Yang and Deb [1] formulated three idealized rules describing the behavior of the Cuckoo species as follows:

- Each cuckoo lays one egg at a time and dumps its egg in a randomly chosen nest.

- The best nests with high quality of eggs will carry over to the next generations.

- The number of available host nests is fixed, and the egg laid by a cuckoo is discovered by the host bird with a probability $p_a \in [0, 1]$. A fraction p_a of the nests is replaced by new ones.

The quality or fitness of a solution can simply be proportional to the value of the objective function. Each egg in a nest represents a solution, and a cuckoo egg represents a new solution; the aim is to use the new and potentially better solutions (cuckoos) to replace a not-so-good solution in the nests [1]. When generating new solutions $X^{(t+1)}$ for cuckoo i, a Lèvy flight is performed as in equation 4.1 [1].

$$X_i^{(t+1)} = X_i^t + \alpha \bigoplus L\grave{e}vy(\lambda) \tag{4.1}$$

where $\alpha > 0$ is the step size which should be related to the scales of the problem of interests. The product $\bigoplus$ means entrywise multiplications. The Lèvy flight essentially provides a random walk while the random step length is drawn from a Lèvy distribution [1]

$$L\grave{e}vy \sim \mu = t^{-\lambda}(1 < \lambda < 3) \tag{4.2}$$

which has an infinite variance with an infinite mean. Here the steps essentially form a random walk process with a power law step-length distribution with a heavy tail. Some of the new solutions should be generated by Lèvy walk around the best solution obtained so far, as this will speed up the local search. However, a substantial fraction of the new solutions should be generated by far-field randomization and whose locations should be far enough from the current best solution. This will make sure the system will not be trapped in a local optimum [1].

CS local search capability can be formulated using equation 4.3 that is used to get new cuckoo solutions based on equation 4.1 [1].

$$X_i^{new} = X_i^{old} + 2 * step * (X_i^{old} - best) \tag{4.3}$$

where *step* is a random number drawn from Lèvy distribution, *best* is the current best solution, X_i^{old} is the old given solution, and X_i^{new} is the newly generated solution.

The explorative power of the CS algorithm is in its strategy for abandoning bad solution where a fraction p_a of solutions is abandoned and new solutions are generated using equation 4.4.

$$X_i^{new} = X_i^{old} + rand1 * (rand2 > p_a) * (X_a - X_b) \tag{4.4}$$

where X_i^{new} is the new nest to be found, new solution, X_i^{old} old nest to be abandoned, $rand1, rand2$ are two random numbers drawn from uniform distribution $\in [0, 1]$, p_a is the probability of a nest being discovered, and X_a, X_b are two randomly selected existing nests.

Based on the above description, the basic steps of the cuckoo search (CS) can be summarized as the pseudo code shown in Algorithm (5).

4.2 Cuckoo search variants

4.2.1 Discrete cuckoo search

A discrete version of the CS algorithm was proposed in [11]. The basic idea is to squash the step size of the CS using sigmoidal activation function 4.5 that is a further threshold to generate a 0 or 1 value using equation 4.6.

begin

- objective function $f(X), X = (x_1, x_2, ..., x_d)^T$
- Generate initial population of n random solutions (nests)
- while (stopping criteria not fulfilled)
    1. Update the cuckoos' position using equation 4.3
    2. Select the best between Cuckoos' solutions and existing nests
    3. Abandon a fraction of solution using equation 4.4
    4. Select the best between abandoned nests and newly generated nests
    5. Update the best solution
- end while

end

Algorithm 5: Pseudo code of the cuckoo search (CS)

$$stp = \frac{1}{1 + \exp^{-\alpha}} \qquad (4.5)$$

where stp is the squashed step size and α is the CS continuous step size.

$$mv = \begin{cases} 1 \text{ if } rand < stp \\ 0 \text{ otherwise} \end{cases} \qquad (4.6)$$

where mv is calculated as in equation 4.6 and rand is a random number drawn from uniform distribution in the range [0,1]. The newly generated discretized solution is calculated as in equation 4.7.

$$x_i^{t+1} = mod(x_i^t + mv, m) + 1 \qquad (4.7)$$

where x_i^{t+1} is the update solution or nest position, x_i^t is the old solution or nest, mv is the discrete step size as calculated in equation 4.6, and m is the range of allowed discrete numbers.

4.2.2 Binary cuckoo search

In the binary CS, the search space is modeled as an n-dimensional boolean lattice, in which the solutions are updated across the corners of a hypercube [5]. In order to build this binary vector, equation 4.8 is used, which can provide only binary values in the boolean lattice restricting the new solutions to only binary values:

$$x_i^j(t+1) = \begin{cases} 1 \text{ if } S(x_i^j(t)) > \sigma \\ 0 \text{ otherwise} \end{cases} \qquad (4.8)$$

where $S(x_i^j(t))$ is calculated as in equation 4.9, $x_i^j(t)$ is the concinnous updated nest position, $x_i^j(t+1)$ is the binary updated binary nest position, and σ $U(0,1)$.

$$S(x_i^j(t)) = \frac{1}{1 + e^{-x_i^j(t)}} \tag{4.9}$$

The BCS is initialized with random binary vectors and uses the same updating methodology for the continuous CS.

4.2.3 Chaotic cuckoo search

Although chaotic search can avoid being caught in local minimum because of its ergodicity, pure chaotic search can obtain good solution only through huge iteration step numbers and it is sensitive to initial solution [4]. A two-stage chaotic CS algorithm is put forward by combining CS algorithm with the chaotic search, in which CS algorithm is used to lead global search and chaos optimization (CO) leads local search according to the result of CS algorithm. In order to maintain population diversity and strengthen the dispersion of the search, the algorithm keeps some superior individuals, dynamically contracts search range in view of the best position of the population, and replaces the worse nest position with the one generated in the contract region randomly. The steps of chaotic cuckoo search algorithm can be described as follows:

- Initialize the population, generate randomly n initialized nests.

- Evaluate solutions and find the best solution/nest.

- Update the nest positions as in CS algorithm; see equation 4.1 except the best solution.

- Compare the new positions and the old ones and keep the better solutions.

- R stands for the possibility that the nest host will recognize the cuckoo egg. Compare it with probability Pa. If $R > Pa$, change the nest position randomly and obtain a set of new nest positions.

- Maintain a set of $1/5$ of the best solutions.

- Perform chaos optimization search to the optimal nest position in the population, and update the new nest position.

- Contract the search space around the best solution.

- Generate the rest $4/5$ nest positions of the population randomly in the contracted space, and evaluate it. If the stopping criterion is not met, return to the second step.

Search region contraction is performed using the following two formulas:

$$x_{ij}^{min} = maxx_{ij}^{min} - r(x_{ij}^{max} - x_{ij}^{min}) \tag{4.10}$$

$$x_{ij}^{max} = minx_{ij}^{max} - r(x_{ij}^{max} - x_{ij}^{min}) \tag{4.11}$$

where j is the current dimension, r is a random number drawn from uniform distribution in the range from 0 to 1, x_{ij}^{max} is the maximum value in dimension j, and x_{ij}^{min} is the minimum value in dimension j.

Logistic equation is used in the chaotic search as indicated in the following equation:

$$y_{n+1,d} = \mu y_{n,d}(1 - y_{n,d}) \tag{4.12}$$

where d is the dimension number, $y_{n,d}$ is the solution to be repositioned, and μ is a constant.

4.2.4 Parallel cuckoo search

A parallel version of the standard cuckoo search is proposed in [6].This parallelization is to try to run more than one population on the same search space and to find a better ratio between exploration and exploitation. This is done by dividing the main population into a number of subpopulations or subgroups. CS algorithm can achieve good results even with a small number of cuckoos. Every sub-flock is running standard CS algorithm on the same search space with a different random seed. After a certain number of generations, the results from all flocks are copied into one array, that array is sorted by the fitness value of the results, and the top quarter of the array is copied back to the flocks. The flocks then continue to execute standard CS algorithm with these results as input. The sub-flocks continue their search from best results from all sub-flocks as a starting point. This method prevents trapping into local optimum.

4.2.5 Cuckoo search for constrained problems

A version of CS is presented in [7] for handling constrained optimization problems. The proposed method uses penalty function to convert the constrained optimization problem into an unconstrained one with modified fitness function. The modified fitness function can be formulated as in the following equation:

$$minF(x) = f(x) + Px, k_j \tag{4.13}$$

where Px, k_j are the penalty functions which can be presented as

$$Px, k_j = \sum_{j=1}^{n_g} k_j.max(0, g_j x^2) \tag{4.14}$$

where the parameters k_j are a suitable constant value, $g_j x$ is the substitution of the vector x in the constraint j, and n_j is the number of constraints.

4.2.6 Cuckoo search with adaptive parameters

The parameters p_a, λ, and α are very critical for the performance of CS; see equation 4.1. In [8], the three aforementioned parameters are adapted throughout the optimization iteration to achieve better performance. The main drawback is using constant values for p_a, and λ appears in the number of iterations to find an optimal solution. If the value of p_a is small and the value of α is large, the performance of the algorithm will be poor and leads to considerable increase in number of iterations. If the value of p_a is large and the value of α is small, the speed of convergence is high but it may be unable to find the best solutions. To improve the performance of the CS algorithm and eliminate the drawbacks with fixed values of p_a and α, the improved cuckoo search (ICS) algorithm uses variables p_a and α. In the early generations, the values of p_a and α must be big enough to enforce the algorithm to increase the diversity of solution vectors. However, these values should be decreased in final generations to result in a better fine-tuning of solution vectors. The values of p_a and α are dynamically changed with the number of generation and expressed in equations 4.15 to 4.17.

$$p_a^t = p_a Max - \frac{t}{NI}(p_a Max - p_a Min) \tag{4.15}$$

$$\alpha^t = \alpha_{max} \exp(c.t) \tag{4.16}$$

$$c = \frac{1}{NI} * \ln \frac{\alpha_{min}}{\alpha_{max}} \tag{4.17}$$

where p_a^t, α^t is the calculated p_a and α at iteration t, NI is the total number of iterations for the optimization, t is the current iteration, and $\alpha_{min}, \alpha_{max}$ are the range for allowed α values.

4.2.7 Gaussian cuckoo search

In [9] a version of CS is making use of Gaussian distribution instead of Lévy distribution as outlined in equation 4.18.

$$x_i^{t+1} = x_i^t + \alpha \oplus \sigma_s \tag{4.18}$$

where x_i^{t+1} and x_i^t are the new and old solutions, α is a constant always 1, and σ_s is a random number drawn from normal distribution using equation 4.19.

$$\sigma_s = \sigma_0 \exp(-\mu k) \tag{4.19}$$

where k is the current generation and $sigma_0$ and μ are constants. Results on sample test function proves good performance for the proposed modification.

Table 4.1: CS hybridizations and sample references

Hybrid with	Target	Sample References
Evolutionary algorithm (EA)	Enhance global searching	[10]
Genetic algorithms (GA)	Exploit crossover and mutation	[11]
Scatter search	Exploit its ability to find the nearest global optimum	[12]
Ant colony optimization	Enhance the convergence speed of ant colony	[13]
Powell search	Exploiting the local searching capability of Powell search	[14]
Pattern search	Exploit pattern search local searching capability	[15, 16]
Bat algorithm	Exploit the local searching (mutation) operator in bat algorithm	[17]
Particle swarm optimization	Exploit information sharing and local experience of PSO	[18]
Levenberg-Marquardt (CSLM)	Exploit local searching capability and speed of Levenberg-Marquardt	[33]
Quantum computing	Exploit the diversity and performance of quantum computing	[24]

4.3 Cuckoo search hybridizations

Exploiting the global searching capability of cuckoo search many hybridizations are performed and stated in the literature. The Table 4.1 outlines samples of the hybridization and the corresponding references.

4.3.1 Cuckoo search hybrid with differential evolution

The standard differential evolution (DE) algorithm is adept at exploring the search space and locating the region of global optimal value, but it is not relatively good at exploiting solution [10]. On the other hand, standard CS algorithm is usually quick at the exploitation of the solution though its exploration ability is relatively poor. A hybrid algorithm is proposed in [10] by integrating DE into CS, so-called DE/CS. The difference between DE/CS and CS is that the mutation and crossover of DE are used to replace the original CS selecting a cuckoo. This method can explore the new search space by the mutation of the DE algorithm and exploit the population information with CS, and therefore can conquer the lack of exploitation of the DE algorithm.

The mutation operator of DE can add diversity of the population to improve the search efficiency. The mutation operator of DE can improve the exploration of the new search space.

Crossover and mutation operators are hired from genetic algorithm (GA) and used with the discrete CS in [11] for solving discrete optimization problems. Two parents' solutions are selected from the current population via roulette wheel selection, then crossover operator between the two parents is performed. Improve the offspring by the mutation operation and insert the resulting offspring to the new population.

4.3.2 Cuckoo search hybrid scatter search

Cuckoo search algorithm has proven its ability in solving some combinatorial problems and finding the nearest global optimum solution in reasonable time and good performance. CS has been used in [12] for the reference set update method in scatter search algorithm to further enhance the obtained intermediate solutions. The improvement provides scatter search with random exploration for search space of problem and more of diversity and intensification for promising solutions based on the cuckoo search algorithm.

4.3.3 Cuckoo search hybrid with ant colony optimization

In [13] a hybrid algorithm is presented between CS and ant colony optimization (ACO). The major disadvantage in the ACO is that while trying to solve the combinatorial optimization problems the search has to be performed much faster, but in ACO ant will walk through the path where the pheromone has been deposited. This acts as if it lures the artificial ants. Hence local search will be performing at the faster rate than in the ACO. In order to overcome this drawback, cuckoo search is used. Hence, CS is used only to allow ants to perform local search and keep the global searching capability of ACO.

4.3.4 Hybrid cuckoo search and Powell search

A modified version of CS with inertia weight hybridized with Powell search is proposed in [14]. The inertia weight is presented to balance the local and global ability of CS. Thus, when generating a new solution x^{t+1} a Lévy flight integrating with inertia weight w_{iter} is performed as

$$x_i^{t+1} = w_{iter}.x_i^t + \alpha \bigoplus L\grave{e}vy(\lambda) \tag{4.20}$$

A reasonable choice for w_{iter} should linearly decrease from a large value to a small value through the course so that the CS algorithm has a better performance compared with fixed w_{iter} settings. The larger w_{iter} has greater global search ability, whereas the small w_{iter} has greater local search ability. The inertia weight w_{iter} is a nonlinear function of the present iteration number ($iter$) at each time step. The proposed adaptation of w_{iter} is given in the following equation:

$$w_{iter} = w_{initial}.\mu^{-iter} \tag{4.21}$$

where $w_{initial}$ is the initial inertia weight value selected in the range [0,1] and μ is a constant value in the range [1.0001,1.005].

Powell search, which is an extension of the basic pattern search method, is based on conjugate direction method and is used to speed up the convergence of nonlinear objective functions. A conjugate direction method minimizes a quadratic function in a finite number function; it can be minimized in a few steps using Powell's method. Powell search is used to locally search around the best solution at each iteration so that better solutions may be obtained.

4.3.5 Hybrid cuckoo search and simplex method

Despite its age, simplex method (SM) is still a method of choice for many practitioners in the fields of statistics, engineering, and the physical and medical sciences because it is easy to code and very easy to use [15]. SM is a fast algorithm to search for a local minimum and applicable for multi-dimensional optimization. It is derivative-free and it converges to minima by forming a simplex and using this simplex to search for its promising directions. A simplex is defined as a geometrical figure which is formed by (N+1) vertices (N: the number of variables of a function). SM is hybridized into the standard CS algorithm for locally applying the SM around the best solution at each iteration. Also, SM is used to enhance the worst solution by conducting SM around it [16].

4.3.6 Cuckoo search hybrid with bat algorithm (BA-CS)

Based on the basic CS algorithm, it draws the bat algorithm into the basic CS algorithm, and makes their respective advantages together, then proposes a hybrid optimization algorithm (BACS) bat algorithm and cuckoo optimization algorithm [17].

Based on the new hybridization the updating of nest position is not directly passed to the next iteration. But it is updated using bat algorithm principles. As in the standard bat algorithm, according to the pulse rate and loudness, the nest's position is updated according to equation 4.22.

$$X_{new} = X_{old} + \epsilon A^t \qquad (4.22)$$

where A^t is the average loudness of all bats at this time, ϵ is a random number uniformally drawn from [-1 1], and X_{new}, X_{old} is the updated and original nest positions.

Finally the algorithm evaluates the fitness values of the bird's nest, and finds out the current optimal position of the bird's nest and the optimal value, and enters the next iteration, continue to search and update position through adopting the basic cuckoo algorithm. The proposed hybridization solves the balance of global search and local search well, thus improving the convergence of the algorithm, and avoiding it falling into local optimum and getting the global optimal solution and enhancing the ability of local optimization and convergent precision at the same time.

4.3.7 Cuckoo search hybrid with particle swarm optimization

In the proposed hybrid algorithm, the ability of communication for cuckoo birds has been added [18]. The goal of this communication is to inform each other about their position and help each other to immigrate to a better place. Each cuckoo bird will record the best personal experience as *pbest* during its own life as well as the global swarm intelligence recorded by the *gbest*. The updated rule for cuckoo s to position as the following:

$$x_{t+1}^i = x_t^i + v_{t+1}^i \tag{4.23}$$

where x_{t+1}^i is the updated nest position, x_t^i is the original nest's position, and v_{t+1}^i is the velocity of change and is calculated as follows:

$$v_{t+1}^i = w_t^i * v_t^i + c_1 * rand * (pbest - x_t^i) + c_2 * rand * (gbest - x_t^i) \tag{4.24}$$

where w is inertia weight which shows the effect of previous velocity vector (v_t^i) on the new vector, c_1 and c_2 are acceleration constants, and $rand$ is a random function in the range $[0, 1]$ and x_i^t is current position of the cuckoo. The communication is actually performed after performing the nest updating of the standard CS.

4.3.8 Cuckoo search hybridized with Levenberg–Marquardt (CSLM)

In [33] a hybridization between cuckoo search and Marquardt optimization is proposed. In that work CS final solution is passed to the Marquardt optimization for further enhancement. Thus, this may be called cascading that exploits the global searching power of CS while exploiting the local searching capability of the Levenberg-Marquardt method.

4.3.9 Cuckoo search hybridized with quantum computing

A quantum inspired cuckoo search (QICSA) which integrates the quantum computing principles such as qubit representation, measure operation and quantum mutation, in the core the cuckoo search algorithm was proposed in [24]. This proposed model focuses on enhancing diversity and the performance of the cuckoo search algorithm.The proposed hybridization contains three essential modules.

The first module contains a quantum representation of cuckoo swarm. The particularity of quantum-inspired cuckoo search algorithm stems from the quantum representation it adopts which allows representing the superposition of all potential solutions for a given problem.

The second module contains the objective function and the selection operator. The selection operator is similar to the elitism strategy used in genetic algorithms [24].

The third module, which is the most important, contains the main quantum cuckoo dynamics. This module is composed of four main operations inspired from quantum computing and cuckoo search algorithm: measurement, mutation, interference, and Lévy flights operations. QICSA uses these operations to evolve the entire swarm through generations [24].

4.4 Cuckoo search in real world applications

This section presents sample applications for cuckoo search. CS is successfully applied in many disciplines in decision support and making, image processing domain, and engineering. The main focus is on identifying the fitness function and the optimization variables used in individual applications. Table 5.2 summarizes sample applications and their corresponding objective(s).

Cuckoo search is applied in [19] for the optimal synthesis of symmetric uniformly spaced linear microstrip antenna arrays. The objective is to determinate the optimal excitations element that produces a synthesized radiation pattern within given bounds specified by a pattern mask. The work in [20] deals with the application of the cuckoo search algorithm in the design of an optimized planar antenna array which ensures high gain, directivity, suppression of side lobes, and increased efficiency and improves other antenna parameters as well.

In [21] CS is used to obtain optimal power flow (OPF) problem solution to determine the optimal settings of control variables. The main objective of the OPF problem is to optimize a chosen objective function such as fuel cost, piecewise quadratic cost function, fuel cost with valve point effects, voltage profile improvement, and voltage stability enhancement, through optimal adjustments of power systems control variables while at the same time satisfying system operating conditions with power flow equations and inequality constraints. CS is used for the active power loss minimization in distribution systems to raise the overall efficiency of power systems. The cuckoo search algorithm is used for minimization of cost that includes capacitor cost and cost due to power loss.

In [22] optimum design of truss structures for both discrete and continuous variables based on the CS algorithm is presented. In order to demonstrate the effectiveness and robustness of the present method, minimum weight design of truss structures is performed and the results of the CS and the selected well-known meta-heuristic search algorithms are compared for both discrete and continuous design of three benchmark truss structures.

A hybrid discrete version of the CS has been used in [11] to tackle the aircraft landing problem (ALP) in order to optimize the usage of existing runways at airports. In [10] a hybrid CS and differential evolution algorithm are proposed. The algorithm targets at finding the safe path by connecting the chosen nodes of the coordinates while avoiding the threat areas and costing minimum fuel and thus helps for uninhabited combat air

Table 4.2: Cuckoo search applications and the corresponding fitness function

Application Name	Fitness Function	Sample References
Linear microstrip antennas array	Determinate the optimal excitations element that produces a synthesized radiation pattern	[19, 20]
Optimal power flow (OPF) problem	Optimize a chosen objective function such as fuel cost, piecewise quadratic cost function, fuel cost with valve point effects, voltage profile improvement, voltage stability enhancement	[21]
Design of truss structures	Minimum weight design of truss structures	[22]
Aircraft landing problem (ALP)	Optimize the usage of existing runways at airports	[10, 11]
Job scheduling problem	Minimize the makespan of performing N jobs	[13, 23]
Knapsack problem (KP)	Choose a subset of items that maximize the knapsack profit, and have a total weight less than or equal to C	[24]
Feature selection	Minimize the scatter function for the points inside each class	[25]
Image segmentation	Intra-cluster variability	[26–29]
Deployment of wireless sensor network	Find the optimal node placement in order to improve the network coverage and connectivity with a minimum coverage hole and overlapping area	[30]
Speaker recognition systems	Maximize matching of sound phonemes	[31]
Train neural network	Minimize network mean square error	[32, 33]
Spam detection	Minimize the total membership invalidation cost of the Bloom filter by finding the optimal false positive rates	[34, 35]
Planar graph coloring	Find the minimum number of colors that can be used to color the regions in a planar map with neighboring regions having different colors	[36]

vehicle (UCAV). Path planning for UCAV is a complicated high dimension optimization problem, which primarily centralizes on optimizing the flight route considering the different kinds of constraints.

In [13] a hybrid CS and ant colony optimization are used to solve the job scheduling problem. In job scheduling we have N jobs and M machines. Each and every job has its own order of execution that has to be performed on M machines. Each job has its own starting time. The objective of this algorithm is to minimize the makespan and it can also be used for job scheduling in scientific and high power computing. In [23] a discrete version of the CS algorithm is used to solve the constrained job scheduling task. The goal of job scheduling is to find best start times for each activity in order to complete the whole project in minimum possible duration time. There are two kinds of constraints, first each activity has some precedence activity and cannot start until precedence activities are finished completely. Second constriction is that the availability of each resource in each period of activity duration is limited. So activity start time might be postponed some-times because of lacking some needed resources or precedence constraints.

Knapsack problem (KP) is a well-known optimization problem. Several problems are encoded like Knapsack problems involving resource distribution, investment decision making, budget controlling, project selection, and so on. This problem was proved to be NP-Hard. A hybrid CS quantum algorithm is proposed in [24] to solve the Knapsack problems. The Knapsack problem can be formulated as having a knapsack with maximum capacity equal to C and a set of N items. Each item i has a profit pi and a weight wi. The problem consists of choosing a subset of items that maximize the knapsack profit, and having a total weight less than or equal to C.

CS is used in feature selection on the face recognition data set in [25]. The CS targets to minimize the scatter function for the points inside each class. The fitness function is calculated as in

$$F = \sqrt{\sum_{i=1}^{L}(E_i - E_o)(E_i - E_o)^t} \tag{4.25}$$

where L is the number of classes.

$$E_o = \frac{1}{k}\sum_{i=1}^{L} k_i E_i \tag{4.26}$$

where k is the number of images in class i.

$$E_i = \frac{1}{k_i}\sum_{j=1}^{k_i} U_{ij}, i = 1, 2, 3, ..L \tag{4.27}$$

where U_{ij}: represents the sample image from class i.

In [26] CS is used to optimally select cluster centers for data clustering problems and proves superior performance in comparison with particle swarm optimizer and genetic optimizer. The k-means centers are localized using the CS in [27]. The aim of the optimization is to minimize k-means fitness function. In this study CS helps k-means to converge to global solution rather than the solutions that may get stuck in local minima in the standard k-means algorithm.

In [28] CS is used with wind-driven optimization to find an optimal threshold set for image data. CS algorithm and wind-driven optimization (WDO) for multilevel thresholding using Kapurs entropy have been employed. For this purpose, the best solution as fitness function is achieved through CS and WDO algorithm using Kapurs entropy for optimal multilevel thresholding. A new approach of CS and WDO algorithm is used for selection of optimal threshold value.

A new metaheuristic scheme based upon CS algorithm for the exudates detection in diabetic retinopathy using multi-level thresholding is presented [29].The proposed method is applied for edge detection and the results obtained by this method were compared with the existing methods.

In wireless sensor network applications with a large-scale area, the sensor nodes are deployed randomly in a noninvasive way. The deployment process will cause some issues such as coverage hole and overlapping that reflect to the performance of coverage area and connectivity. Node placement model is constructed to find the optimal node placement in [30]. Virtual force algorithm (VFA) and cuckoo search (CS) algorithm approach for node placement technique is analyzed to find the optimal node placement in order to improve the network coverage and connectivity with a minimum coverage hole and overlapping area.

In [31] CS has been applied on speaker recognition systems and voice. The process of speaker recognition is optimized by a fitness function that matches voices being done on only the extracted optimized features produced by the CS algorithm. CS is used to train a feed-forward neural network in [32]. The CS can quickly find optimal selection for weight/bias satisfying a given error function and avoiding the local minima that always are a problem in gradient-based methods.

A method for training a feed-forward neural network based on CS is proposed in [33]. The proposed method makes use of CS to find the optimal weight/bias values to maximize a given error function. The resulting weights and biases are further enhanced using the Levenberg-Marquardt backpropagation algorithm.

In [34] a system for spam detection based on CS is proposed. A spam word is a list of well-known words appearing in spam mails. Bin Bloom Filter (BBF) groups the words into a number of bins with different false-positive rates based on the weights of the spam words. A CS algorithm is employed to minimize the total membership invalidation cost of the BFs by finding the optimal false-positive rates and number of elements stored in every bin. In [35] authors propose the application of cuckoo search optimization algorithm in web document clustering area to locate the optimal centroids of the cluster and to find global solution of the clustering algorithm. Web document clustering helps in the development of new techniques for helping users to effectively navigate, summarize, and organize the overwhelming information to ease the process of finding the relevant information on the Web.

The minimum number of colors that can be used to color the regions in a planar map with neighboring regions having different colors has been a problem of interest for over a century.In [36] an improved cuckoo search optimization (ICS) algorithm is proposed for solving the planar graph coloring problem. The improved cuckoo search optimization algorithm consists of the walking one strategy, swap and inversion strategy, and greedy strategy. The proposed improved cuckoo search optimization algorithm can solve the planar graph coloring problem using four colors more efficiently and accurately.

4.4.1 Cuckoo search in feature selection

In this case study CS optimizer is used to optimally select a subset of feature to maximize classification performance. Ten datasets chosen from the UCI machine learning repository [18] are used in the experiments and comparison results; see Table 4.3. The ten datasets were selected to have various numbers of attributes and instances as representatives of various kinds of issues that the proposed technique would be tested on. For each dataset, the instances are randomly divided into three sets, namely, training, validation, and testing sets in cross-validation manner.

The well-known K-NN is used as a classifier to evaluate the final classification performance for individual algorithms with k = 5 [17]. Each optimization algorithm is run for M times to test convergence capability for an optimizer. The indicators used to compare the different algorithms are

Table 4.3: Description of the data used in the experiments

Dataset	No. of Features	No. of Samples
WineEW	13	178
spectEW	22	267
sonarEW	60	208
penglungEW	325	73
ionosphereEW	34	351
heartEW	13	270
congressEW	16	435
breastEW	30	569
krvskpEW	36	3196
waveformEW	40	5000

- *Classification average accuracy*: This indicator describes how accurate the classifier is given the selected feature set. The classification average accuracy can be formulated as

$$AvgPerf = \frac{1}{M} \sum_{j=1}^{M} \frac{1}{N} \sum_{i=1}^{N} Match(C_i, L_i) \tag{4.28}$$

where M is the number of times to run the optimization algorithm to select feature subset, N is the number of points in the test set, C_i is the classifier output label for data point i, L_i is the reference class label for data point i, and $Match$ is a function that outputs 1 when the two input labels are the same and outputs 0 when they are different.

- *Statistical mean* is the average of solutions acquired from running an optimization algorithm for different M running. Mean represents the average performance a given stochastic optimizer can be formulated as.

$$Mean = \frac{1}{M} \sum_{i=1}^{M} g_*^i \tag{4.29}$$

where M is the number of times to run the optimization algorithm to select feature subset, and g_*^i is the optimal solution resulted from run number i.

- *Std* is a representation for the variation of the obtained best solutions found for running a stochastic optimizer for M different runs. Std is used as an indicator for optimizer stability and robustness where Std is smaller; this means that the optimizer converges always to the same solution, while larger values for Std mean much random results. Std is formulated as

$$Std = \sqrt{\frac{1}{M-1} \sum (g_*^i - Mean)^2} \tag{4.30}$$

where M is the number of times to run the optimization algorithm to select feature subset, g_*^i is the optimal solution resulted from run number i, and $Mean$ is the average defined in equation 4.29.

- *Average selection size* represents the average size of the selected features to the total number of features. This measure can be formulated as

$$AVGSelectionSZ = \frac{1}{M} \sum_{i=1}^{M} \frac{size(g_*^i)}{D} \tag{4.31}$$

where M is the number of times to run the optimization algorithm to select feature subset, g_*^i is the optimal solution resulted from run number i, $size(x)$ is the number of values for the vector x, and D is the number of features in the original dataset.

Table 4.4: CS parameter setting

Parameter	*Value*	*Meaning*
NIter	80	Total number of iterations used
n	10	Number of search agents
pa	0.25	Nest abandon probability

CS optimizer is evaluated against other common optimizers such as the particle swarm optimization (PSO) method proposed in [53] and the genetic algorithm proposed in [52] for optimal searching of the feature space.

The parameter setting used for the different optimizers are in Table 4.4.

Each algorithm is run for $M = 10$ times with new random initialization and best feature combination selected.

Table 4.5 displays the average of the fitness function, statistical mean, of the best solution found by the different optimization algorithms at the different datasets. This indicator proves the capability of optimization algorithm to find optimum solution. We can see from the table that CS finds better solutions than PSO and GA, which proves its searching capability.

Table 4.6 outlines the average performance of the selected features by the different algorithms over the test data. We can remark that the CS performs better on the test data than GA and PSO.

Table 4.7 outlines the variation of the obtained best solutions for the different runs of each optimizer on the different dataset. We can see that the minimum variation of output fitness value is achieved by CS, which proves that the CS always converges to optimal/near-optimal solution regardless of the random initialization of the algorithm.

Table 4.5: Experimental results of statistical mean of fitness function of the best solution for the different methods

Dataset	*CS*	*GA*	*PSO*
breastEW	0.014737	0.029474	0.026316
congressEW	0.024828	0.042759	0.044138
heartEW	0.133333	0.14	0.155556
ionosphereEW	0.083761	0.109402	0.11453
krvskpEW	.026992	0.047048	0.075726
penglungEW	0.191667	0.208333	0.2
sonarEW	0.118841	0.188406	0.165217
spectEW	0.110112	0.139326	0.137079
waveformEW	0.185868	0.215329	0.21018
WineEW	0.00678	0.016949	0.023729

Table 4.6: Performance of the different algorithms on test data

Dataset	*CS*	*GA*	*PSO*
breastEW	0.942105	0.944211	0.94
congressEW	0.947586	0.931034	0.935172
heartEW	0.811111	0.793333	0.753333
ionosphereEW	0.866667	0.818803	0.85812
krvskpEW	0.96939	0.945352	0.913427
penglungEW	0.68	0.672	0.64
sonarEW	0.737143	0.645714	0.68
spectEW	0.806742	0.795506	0.802247
waveformEW	0.792192	0.764805	0.768168
WineEW	0.943333	0.926667	0.93

Table 4.7: Standard deviation of fitness functions for the different optimizers

Dataset	*CS*	*GA*	*PSO*
breastEW	0.005766	0.007061	0.007443
congressEW	0.007863	0.00577	0.012528
heartEW	0.028328	0.030021	0.028328
ionosphereEW	0.014044	0.020406	0.018726
krvskpEW	0.007824	0.007597	0.050853
penglungEW	0.095924	0.10623	0.090331
sonarEW	0.036087	0.043478	0.040476
spectEW	0.033141	0.03237	0.018463
waveformEW	0.009586	0.015021	0.009286
WineEW	0.01516	0.011985	0.025705

Table 5.6 outlines the selected feature size from the different optimizers on the different datasets. We can see that the CS has comparable feature sizes to both GA and PSO, while it keeps better classification accuracy as we outlined in Table 4.6.

4.4.2 A modified cuckoo search for solving a convex economic dispatch problem

Economic load dispatch (ELD) is a non-linear global optimization problem for determining the power shared among the generating units to satisfy the generation limit constraints of each unit and minimizing the cost of power production. Cuckoo search is one of the most powerful swarm intelligence algorithms and it has been applied to solve global optimization problems; however, it suffers from slow convergence and stagnation. The economic load dispatch (ELD) problem can be mathematically formulated as a continuous optimization problem. The goal of the economic dispatch problem is to find the

Table 4.8: Mean selection size for the different optimizers over the different datasets

Dataset	CS	GA	PSO
breastEW	0.593333	0.613333	0.52
congressEW	0.425	0.6	0.475
heartEW	0.646154	0.553846	0.461538
ionosphereEW	0.435294	0.494118	0.447059
krvskpEW	0.483333	0.488889	0.566667
penglungEW	0.511385	0.486154	0.473846
sonarEW	0.473333	0.553333	0.51
spectEW	0.554545	0.536364	0.472727
waveformEW	0.535	0.605	0.57
WineEW	0.553846	0.538462	0.507692

optimal combination of power generation in such a way that the total production cost of the system is minimized. The cost function of a generating unit can be represented as a quadratic function with a sine component. The sine component denotes the effect of steam valve operation. The quadratic refers to a convex objective, whereas the valve-point effect makes the problem non-convex. The convex and non-convex cost function of a generator can be expressed as follows:

$$F_i(P_i) = a_i P_i^2 + b_i P_i + c_i \qquad (4.32)$$

$$F_i(P_i) = a_i P_i^2 + b_i P_i + c_i + |d_i sin[e_i * (P_i^{min} - P_i)]| \qquad (4.33)$$

where P_i is the active power output, $F_i(P_i)$ is the generation cost, P_i^{min} is the minimum output limit of the generator, and a_i, b_i, c_i, d_i, e_i are the cost coefficients of the generator. The fuel cost of all generators can be defined by the following equation:

$$MinF_i(P_i) = \sum_i^{Ng} a_i P_i^2 + b_i P_i + c_i + |d_i sin[e_i * (P_i^{min} - P_i)]| \qquad (4.34)$$

where Ng is the number of generating units.

In order to investigate the efficiency of the modified cuckoo search (MCS) algorithm, we present the general performance of it with a six-generator test system and the parameter setting of the algorithm is reported in Table 4.9.

Table 4.9: Parameter setting

Parameters	Definitions	Values
N	Population size	40
p_a	A fraction of worse nests	0.25
Max_{itr}	Maximum number of iterations	500

Table 4.10: Generator active power limits

Generator	1	2	3	4	5	6
$P_{min}(MW)$	10	10	35	35	130	125
$P_{max}(MW)$	125	150	225	210	325	315

Table 4.11: Fuel cost coefficients

No.	a	b	c
1	0.15240	38.53973	756.79886
2	0.10587	46.15916	451.32513
3	0.02803	40.39655	1049.9977
4	0.03546	38.30553	1243.5311
5	0.02111	36.32782	1658.5596
6	0.01799	38.27041	1356.6592

The MCS algorithm is tested on a six-generator test system. In order to simplify the problem, the values of parameters d, e in Equation 4.33 have been set to zero. The proposed algorithm is carried out for a total demand of 700 MW and 800 MW. The generator active power limits and the fuel cost coefficients are given in Tables 4.10 and 4.11, respectively.

The general performance of the MCS algorithm with economic dispatch problem is shown in Figure 4.1 by plotting the number of iterations versus the cost ($/h$) for total system demand 700 MW and 800 MW, respectively.

Figure 4.1 shows that the cost values are rapidly decreasing with a few numbers of iterations. We can conclude from Figure 4.1 that the algorithm is promising and can obtain the desired power with minimum cost.

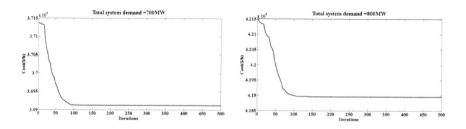

Figure 4.1: The general performance of MCS with economic dispatch problem

4.5 Chapter conclusion

This chapter reviewed the cuckoo search algorithm, its behavior as well as its variants. A hybridization of the cuckoo search with other approaches was discussed. We gave an overview of real life application as well as results of two cases in feature selection and a modified version cuckoo search for solving a convex economic dispatch problem.

Bibliography

[1] Xin-She Yang Suash Deb, "Cuckoo Search via Levy Flights", 2009 World Congress on Nature and Biologically Inspired Computing (NaBIC 2009), USA, pp. 210-214, 2009.

[2] X.-S. Yang and S. Deb, Engineering optimisation by cuckoo search, Int. J. Mathematical Modelling and Numerical Optimisation, Vol. 1(4), pp. 330-343, 2010.

[3] Hongqing Zheng, Yongquan Zhou and Peigang Guo, "Hybrid genetic-cuckoo search algorithm for solving runway dependent aircraft landing problem", *Research Journal of Applied Sciences, Engineering and Technology* 6(12), pp. 2136-2140, 2013.

[4] Aijia Ouyang, Guo Pan, Guangxue Yue, and Jiayi Du, "Chaotic cuckoo search algorithm for high dimensional functions", *Journal of Computers*, vol. 9(5), pp. 1282-1290, MAY 2014.

[5] Xin-She Yang, Cuckoo search and firefly algorithm, Vol 516, pp. 146-148.

[6] Milos Subotic, Milan Tuba, Nebojsa Bacanin, Dana Simian, "Parallelized cuckoo search algorithm for unconstrained optimization", Proceeding BICA'12 Proceedings of the 5th WSEAS Congress on Applied Computing Conference, and Proceedings of the 1st International Conference on Biologically Inspired Computation, pp. 151-156, 2014.

[7] Radovan R. Bulatovi, Goran Bokovi, Mile M. Savkovi, and Milomir M. Gai, "Improved cuckoo search (ICS) algorithm for constrained optimization problems", *Latin American Journal of Solids and Structures*, vol.11(8), pp. 1349-1362, 2014.

[8] Ehsan Valian, Shahram Mohanna, and Saeed Tavakoli, "Improved cuckoo search algorithm for global optimization", *International Journal of Communications and Information Technology, IJCIT*, vol.1(1), pp. 31-44, Dec. 2011.

[9] Hongqing Zheng and Yongquan Zhou, "A novel cuckoo search optimization algorithm base on gauss distribution", *Journal of Computational Information Systems*, Vol. 8(10), pp. 4193-4200, 2012.

[10] Gaige Wang, Lihong Guo, Hong Duan, Luo Liu, Heqi Wang, Jianbo Wang, "A hybrid meta-heuristic DE/CS algorithm for UCAV path planning", *Journal of Information and Computational Science*, vol. 9(16), pp. 4811-4818, 2012.

[11] Hongqing Zheng, Yongquan Zhou, and Peigang Guo, "Hybrid genetic-cuckoo search algorithm for solving runway dependent aircraft landing problem", *Research Journal of Applied Sciences, Engineering and Technology*, Vol. 6(12), pp. 2136-2140, 2013

[12] Ahmed T. Sadiq Al-Obaidi, "Improved scatter search using cuckoo search, (IJARAI)" *International Journal of Advanced Research in Artificial Intelligence*, Vol. 2(2), pp. 61-67, 2013.

[13] R.G. Babukartik and P. Dhavachelvan, "Hybrid algorithm using the advantage of ACO and cuckoo search for job scheduling", *International Journal of Information Technology Convergence and Services (IJITCS)*, Vol. 2(4), August 2012.

[14] Wen Long and Jianjun Jiao, "Hybrid cuckoo search algorithm based on powell search for constrained engineering design optimization", *WSEAS Transaction on Math.*, Vol. 13, pp. 431-440, 2014.

[15] Hongqing Zheng, Qifang Luo, Yongquan Zhou, "A novel hybrid cuckoo search algorithm based on simplex operator", *International Journal of Digital Content Technology and its Applications (JDCTA)*, Vol. 6(13), pp. 45-52, July 2012.

[16] Wen Long, Wen-Zhuan Zhang, Ya-Fei Huang, and Yi-xiong Chen, "A hybrid cuckoo search algorithm with feasibility-based rule for constrained structural optimization", *J. Cent. South Univ.*, Vol. 21, pp. 3197-3204, 2014.

[17] Yanming Duan, "A hybrid optimization algorithm based on bat and cuckoo search", *Advanced Materials Research*, Vols. 926-930, pp. 2889-2892, 2014.

[18] Amirhossein Ghodrati and Shahriar Lotfi, "A hybrid CS/PSO algorithm for global optimization", Intelligent Information and Database Systems Lecture Notes in Computer Science, Vol. 7198, pp. 89-98, 2012.

[19] Haffane Ahmed and Hasni Abdelhafid, "Cuckoo search optimization for linear antenna arrays synthesis", *Serbian Journal of Electrical Engineering*, Vol. 10(3), pp. 371-380, October 2013.

[20] A. Sai Charan, N.K. Manasa, and N.V.S.N. Sarma, "Uniformly spaced planar array optimization using cuckoo search space algorithm", *Computer Science and Information Technology (CS and IT)*, pp. 157-167, IT-CSCP 2014.

[21] M. RamaMohana Rao and A.V. Naresh Babu, "Optimal power flow using cuckoo optimization algorithm", *International Journal of Advanced Research in Electrical, Electronics and Instrumentation Engineering*, Vol. 2(9), pp.4213-4218, September 2013.

[22] A. Kaveh and T. Bakhshpoori, "Optimum design of space trusses using cuckoo search algorithm with Levy flights", *Transactions of Civil Engineering*, Vol. 37(C1), pp. 1-15, 2013.

[23] Mehdi Hampaeyan Miandoab, "A new discrete cuckoo search for resource constrained project scheduling problem (rcPSP)", *International Journal of Advanced Studies in Humanities and Social Science* Vol. 1(11), pp. 2368-2378, 2013.

[24] Abdesslem Layeb, "A novel quantum inspired cuckoo search for knapsack problems", *Journal International Journal of Bio-Inspired Computation Archive*, Vol. 3(5), pp. 297-305, September 2011.

[25] Vipinkumar Tiwari, "Face recognition based on cuckoo search algorithm", *Indian Journal of Computer Science and Engineering (IJCSE)*, Vol. 3(3), pp. 401-405, June-July 2012.

[26] J. Senthilnath, Vipul Das, S.N. Omkar, V. Mani, "Clustering Using Levy Flight Cuckoo Search", Proceedings of Seventh International Conference on Bio-Inspired Computing: Theories and Applications (BIC-TA 2012), *Advances in Intelligent Systems and Computing 202*, Springer India 2013.

[27] Rui Tang, Simon Fong, Xin-She Yang, and Suash Deb, "Integrating Nature-inspired Optimization Algorithms to K-means Clustering", 2012 Seventh International Conference on Digital Information Management (ICDIM), 22-24 Aug. Macau, pp. 116-123, 2012.

[28] Ashish Kumar Bhandari, Vineet Kumar Singh, Anil Kumar, Girish Kumar Singh, "Cuckoo search algorithm and wind driven optimization based study of satellite image segmentation for multilevel thresholding using Kapur entropy", *Expert Systems with Applications*, Vol. 41, pp. 3538-3560, 2014.

[29] Srishti, "Technique based on cuckoo's search algorithm for exudates detection in diabetic retinopathy", *Ophthalmology Research: An International Journal*, Vol. 2(1), pp. 43-54, 2014.

[30] Puteri Azwa Ahmad, M. Mahmuddin, and Mohd Hasbullah Omar, "Virtual Force Algorithm And Cuckoo Search Algorithm For Node Placement Technique In Wireless Sensor Network", Proceedings of the 4th International Conference on Computing and Informatics, ICOCI 2013, 28-30 August, 2013 Sarawak, Malaysia. Universiti Utara Malaysia.

[31] Monica Sood and Gurline Kaur, "Speaker recognition based on cuckoo search algorithm", *International Journal of Innovative Technology and Exploring Engineering (IJITEE)*, Vol. 2(5), pp. 311-313, April 2013.

[32] N.M. Nawi, A. Khan, and M.Z. Rehman, "A New Back-Propagation Neural Network Optimized with Cuckoo Search Algorithm", (ICCSA2013), Part I, LNCS 7971, pp. 413-426, 2013.

[33] Nazri Mohd. Nawi, Abdullah Khan, and Mohammad Zubair Rehman, "A New Cuckoo Search Based Levenberg-Marquardt (CSLM) Algorithm", ICCSA 2013, Part I, LNCS 7971, pp. 438-451, Springer-Verlag Berlin Heidelberg, 2013

[34] Arulanand Natarajan, S. Subramanian, K. Premalatha, "An enhanced cuckoo search for optimization of bloom filter in spam filtering", *Global Journal of Computer Science and Technology* Volume XII Issue I Version I, Online ISSN: 0975-4172, (2012).

[35] Moe Moe Zaw and Ei Ei Mon, "Web document clustering using cuckoo search clustering algorithm based on Levy flight", *International Journal of Innovation and Applied Studies*, Vol. 4(1), pp. 182-188, Sept. 2013.

[36] Yongquan Zhou, Hongqing Zheng, Qifang Luo, and Jinzhao Wu, "An improved cuckoo search algorithm for solving planar graph coloring problem", *Applied Mathematics and Information Sciences*, Vol. 7(2), pp. 785-792, 2013.

5 Firefly Algorithm (FFA)

5.1 Firefly algorithm (FFA)

FA is a nature-inspired stochastic global optimization method that was developed by Yang [1]. This algorithm imitates the mechanism of firefly mating and exchange of information using light flashes. This chapter presents the main behavior of fireflies, the artificial FFA, and the variants added to the basic algorithm that was proposed in 2008. Sample applications of FFA are presented at the end of the chapter.

5.1.1 Behavior of fireflies

There are about two thousand firefly species, and most fireflies produce short and rhythmic flashes [2]. The fundamental functions of such flashes are:

- Attraction of mating partners (communication)
- Attraction of potential prey
- Provides warning mechanism

Two combined factors make most fireflies visible only to a limited distance [2]. The first is that the light intensity at a particular distance r from the light source obeys the inverse square law. That is to say, the light intensity I decreases as the distance r increases in terms of $I \propto \frac{1}{r^2}$, i.e., light intensity is inversely proportional to the square of the distance. The second is that the air absorbs light which becomes weaker and weaker as the distance increases.

The a forementioned manners and rules are formulated mathematically in the artificial firefly to be discussed in the next section.

5.1.2 Artificial fireflies

Yang in [2] formulated three idealized rules that describe the behavior of artificial fireflies as follows:

- All fireflies are unisex so that one firefly will be attracted to other fireflies regardless of their sex;

- Attractiveness is proportional to the brightness, thus for any two flashing fireflies, the less brighter one will move toward the brighter one. The attractiveness decreases as the distance increases between two fireflies. If there is no brighter one than a particular firefly, it will move randomly;

- The brightness of a firefly is affected or determined by the landscape of the objective function. For a maximization problem, the brightness can simply be proportional to the value of the objective function.

The movement of a firefly i attracted to another more attractive (brighter) firefly j is determined by

$$x_i = x_i + \beta_0 e^{-\gamma r_{ij}^2}(x_j - x_i) + \alpha(rand - 0.5) \tag{5.1}$$

where the second term is due to the attraction while the third term is randomization with α being the randomization parameter. $rand$ is a random number drawn from uniform distribution in $[0, 1]$ and hence the expression $(rand - 0.5)$ ranges from [-0.5,0.5] to allow for positive and negative variation. β_0 is always set to 1 and $\alpha \in [0, 1]$. The α parameter physically represents the noise existing in the environment and affects the light transmission, while in the artificial algorithm it can be selected to allow for solution variation and hence provide for more diversity of solutions. Furthermore, the randomization term can easily be extended to a normal distribution with 0 mean and variance $1; N(0, 1)$ to allow for variation of noise rate in the environment.

The parameter γ characterizes the variation of the attractiveness, and its value is crucially important in determining the speed of the convergence and how the FA algorithm behaves. In most applications, it typically varies from 0.01 to 100. The distance between $Firefly_i, Firefly_j$ is denoted as r_{ij} and defined in equation 5.2.

$$r_{ij} = \|x_i - x_j\| \tag{5.2}$$

where x_i represents the position of firefly i.

Remark that the attractiveness coefficient in the updating equation $\beta_0 e^{-\gamma r_{ij}^2}$ is used as an approximation of the light intensity loss by distance as mentioned in the idealized rules. It can be modeled by any monotonically decreasing functions. Also, the random term in the equation is used to formulate the effect of dust and environment on the light intensity. So, the firefly algorithm (FFA) which is inspired by the behavior of fireflies can be formulated in the following pseudo-code; see Algorithm 6.

For more details about the implementation reader can refer to [3]. FFA properties can be formulated in the following items:

- FFA is a swarm intelligent method that has the advantages of swarm optimization.

- FFA can easily handle multi-modal problems thanks to its automatic subdivision of population, where the scope vision of each firefly is limited which allows fireflies to make sub-swarms in the search space.

- The randomness and attraction parameters of the FFA can be easily tuned throughout the iteration to enhance the convergence speed of the algorithm.

input : n Number of Fireflies
$NIter$ Number of iterations for optimization
γ attractiveness parameter
α contribution of the random term (environment noise)
output: Optimal firefly position and its fitness

Initialize a population of n fireflies' positions at random
Find the best solution based on fitness;
while *Stopping criteria not met* **do**
  foreach $Firefly_i$ **do**
  foreach $Firefly_j$ **do**
  if *firefly j is better than firefly i* **then**
  | Move firefly i towards firefly j using equation 5.1
  else

  end
  Evaluate The positions of individual fireflies.
end

Algorithm 6: Pseudo code of the firefly algorithm

5.2 FFA variant

5.2.1 Discrete FFA

A discrete version of the standard continuous FFA is proposed in [29]. The proposed algorithm can be formulated as digitization of the *initialization, distance, attraction,* and *random* component in the standard continuous FFA updating equation; see equation 5.1. These steps can be displayed in some detail as follows:

- *Initial Fireflies:* Random permutations are generated an equal number of fireflies in the required space $\Re^n$.

- *Distance Function:* Hamming distance is used where it counts the number of non-corresponding elements in the sequence. The Hamming distance between two permutations is the number of non-corresponding elements in the sequence.

Example: consider the following three fireflies with positions

$$\pi_1 = [1, 2, 3, 4, 5, 6] \tag{5.3}$$

$$\pi_2 = [1, 2, 4, 3, 6, 5] \tag{5.4}$$

$$\pi_3 = [1, 2, 4, 5, 6, 3] \tag{5.5}$$

The hamming distance between π_1, π_2 is 4, while the Hamming distance between π_2, π_3 is 2, as there are 4 corresponding elements between π_2, π_3.

- *Attraction Step:* Attraction is the first component of the firefly updating equation. In the discrete version a firefly π_1 is attracted to firefly π_2, denoted as $\pi_{1\rightarrow2}$. First, the common corresponding elements are set in the output solution. Second, elements are copied from solution π_1 to the output solution with probability:

$$\beta = \frac{1}{1 + \gamma d_{\pi1,\pi2}} \tag{5.6}$$

  where $d_{\pi1,\pi2}$ is the Hamming distance between the two fireflies and the parameter γ characterizes the variation of the attractiveness as defined above in the continuous version. The other elements are retained from π_2 solution. While doing this, we need to watch not to create duplicate elements in the resulting permutation. Duplicate elements are filled with random unused numbers.

  The attraction step can be formulated as in the following equation:

$$x_i, d^{t+1} = \begin{cases} x_{i,d}^t & \text{if } x_{i,d}^t = x_{j,d}^t \\ x_{j,d}^t & \text{if } r1 < \beta \\ x_{i,d}^t & \text{if } r1 \geq \beta \text{ and } x_{i,d}^t \text{ not previously appeared in} x_{i,d}^t \end{cases} \tag{5.7}$$

  where

  - $r1$ is a random number generated from uniform distribution in the range [0,1].
  - $x_{i,d}^t$, is the value of solution i in dimension d at time t.
  - β is attraction coefficient calculated as in equation 5.6.

  The result of equation 5.7 is some missing values in some dimensions that further will be filled with random allowed numbers with no repetition.

- *Random Component:* It should allow us to shift the permutation into one of the neighboring permutations. There are two ways how to apply the α-step: either to make an $\alpha.Random()$ many swaps of randomly chosen two elements, or to choose an $\alpha.Random()$ many elements, and shuffle their positions.

5.2.2 Binary FFA

A binary version of the standard firefly algorithm was proposed in [6]. The algorithm forces the firefly positions to be on the binary grid. The FFA standard updating equation (see equation 5.1) allows updated the binary FFA position which may lead to continuous values. The continuous values are mapped into binary values by means of a probabilistic rule based on a sigmoidal transformation applied to each component of the real-valued position vector as shown in the following equation:

$$X_{id} = \begin{cases} 1 & if rand < \frac{1}{1+exp(-x_{id})} \\ 0 & otherwise \end{cases} \tag{5.8}$$

where X_{id} is the position of the ith agent in dimension d and $rand$ is a random number drawn from uniform distribution in the range from 0 to 1.

The aforementioned equation forces the firefly position/solutions to be $\in 0, 1$. The same binarization step is used in [7] but using the hyperbolic tan function for squashing the continuous values before threshold; see next equation.

$$tanh(x_p) = \frac{exp(2 * |x_p|) - 1}{exp(2 * |x_p|) + 1} \tag{5.9}$$

where x_p is the continuous value of the solution x in dimension d. The final binary value for the discrete solution can be calculated as

$$X_{id} = \begin{cases} 1 & if \text{ rand } < tanh(x_p) \\ 0 & otherwise \end{cases} \tag{5.10}$$

where $tanh$ is the hyperbolic tan function.

5.2.3 Chaotic firefly algorithm

The rate of convergence of FA is strongly influenced by the choice of algorithm parameters [8]. The success and rapidness of the search is highly dependent on a good trade-off between the exploration and the exploitation. Exploration allows search of the entire search space by ensuring the redirection of the search toward new regions, while exploitation favors a quick convergence toward the optimum. Santos Coelho [9] pointed out that use of chaotic sequences can help the FA to escape more easily from local minima than through traditional FA. A logistic map was used to tune the randomization parameter (α) and the attraction variation parameter (γ) to generate the novel chaotic firefly algorithm (CFA).

The updating for the α and γ parameters are adapted according to the following:

$$\gamma^{t+1} = \mu_1 \gamma^t [1 - \gamma^t] \tag{5.11}$$

$$\alpha^{t+1} = \mu_2 \alpha^t [1 - \alpha^t] \tag{5.12}$$

where μ_1, μ_2 are constants of the logistic chaos function, $alpha^t$ is the firefly attraction variation parameter at iteration t, and γ^t is the firefly randomization parameter at iteration t.

5.2.4 Parallel firefly

Motivated by advance in multi-core technology a parallel version of FFA is presented in [10]. Parallelization of algorithms has been proven to be a very powerful method in the case of population-based algorithms. Multiple colony approach is used in this work for parallelizing the FFA as follow.

- Step 1 (Initialization): Set the number of colonies/subpopulation each with its own fireflies on an individual threads and define the communication time; i.e., for every 100 iterations.

- Step 2 (Execution): Run firefly optimization on individual threads.

- Step 3 (Communication): Sort agents in individual threads according to their attractiveness and divide them into two groups. Every subpopulation exchanges half its agents with another subpopulation.

After that, sub-colonies are replaced with a new population and search process is continued.

5.2.5 FFA for constrained problems

A version of FFA to handle optimization problems with constraints is proposed in [11]. The constrained problem is transformed into an unconstrained one by forming a pseudo-objective function as follows:

$$\Phi(x, r) = f(x) + r\{\sum_{i=1}^{p}[h_i(x)]^2 + \sum_{i=1}^{m}[g_i^+(x)]^2\} \tag{5.13}$$

where $g_i^+(x) = max(0, g_i(x))$, $g_i(x)$ represents the inequality constraints where $g_i(x) \leq 0$, $h_i(x)$ represents the equality constraints and r is the scalar penalty parameter. Another propose in the same paper is based on a set of rules that helps FFA deal with constraints as follows:

- If both fireflies are at feasible positions and firefly j is at a better position than firefly i, then firefly i moves toward firefly j.

- If firefly i is at an infeasible position and firefly j is at a feasible position, then i moves to firefly j.

- If positions of firefly i and firefly j are infeasible and the number of constraints satisfied by firefly j is more than that of firefly i, then firefly i moves to firefly j.

- Once the position of the firefly is updated using above rules 1 to 3, if the updated position of the firefly i presents improved solution over the solution associated with its previous iteration position, then firefly i accepts its current solution, otherwise retains its previous iteration solution.

5.2.6 Lèvy flight FFA (LFA)

Numerical studies and results suggest that the proposed Lévy-flight firefly algorithm is superior to existing metaheuristic algorithms and and motivates using this distribution in combination with the standard FFA [12]. The random component in the standard FFA is replaced by the Lèvy flight rather than the uniform random numbers used in the standard FFA; see equation 5.1. The movement of a firefly i attracted to another more attractive (brighter) firefly j is determined by

$$x_i = x_i + \beta_0 e^{-\gamma r_{ij}^2}(x_j - x_i) + \alpha sign(rand - 0.5) \bigoplus L\grave{e}vy \qquad (5.14)$$

where the second term is due to the attraction while the third term is randomization with α being the randomization parameter. $rand$ is a random number generated uniformly in $[0, 1]$. β_0 is always set to 1 and $\alpha \in [0, 1]$. The product $\bigoplus$ means entrywise multiplications. The $sign(rand - 0.5)$ where $rand \in [0, 1]$ provides a random sign or direction while the random step length is drawn from a Lèvy distribution.

$$L\grave{e}vy \ \mu = t^{-\lambda}, (1 < \lambda \leq 3) \qquad (5.15)$$

which has an infinite variance with an infinite mean. Here the steps of firefly motion are essentially a random walk process with a power-law step-length distribution with a heavy tail.

5.2.7 Intelligent firefly algorithm (IFA)

IFA is proposed in [13] for the global optimization. In the original FA, the movement of a firefly is determined mainly by the attractiveness of the other fireflies. A firefly can be attracted to another firefly merely because it is close to it even if it is not so much better, which may take it away from the global minimum that may be far in distance from it. A firefly is pulled toward other fireflies as each of them contributes to the move by its attractiveness. This behavior may lead to a delay in the collective move toward the global minimum. In the intelligent firefly algorithm (IFA) the ranking is performed such that the firefly is attracted only as a subset ϕ of the better fireflies. This fraction represents a top portion of the fireflies based on their rank. Thus, a firefly is acting intelligently by basing its move on the top ranking fireflies only and not merely on attractiveness.

The strength of FA is that the location of the best firefly does not influence the direction of the search. Thus, the fireflies are not trapped in a local minimum. However, the search for the global minimum requires additional computational effort as many fireflies wander around uninteresting areas. With the intelligent firefly the parameter ϕ can maintain the advantage of not being trapped in a local minimum while speeding up the search for the global minimum. The value of ϕ gives a balance between the ability of the algorithm not to be trapped in a local minimum and its ability to exploit the best solutions found [13].

5.2.8 Gaussian firefly algorithm (GDFF)

The GDFF proposed in [14] applies three behaviors to improve performance of the firefly algorithm.

1. *Adaptive step size:* All fireflies move with a fixed length in all iterations in the standard FFA. The algorithm may miss better local search capabilities and sometimes it traps into several local optimums. In the proposed algorithm, it is defined as weight for α that depends on iterations and it always produces a value less than one. This coefficient is determined by equation 5.16.

$$\alpha_i = \alpha_{min} + \frac{(iter_{max} - i)^n}{iter_{max}^n} * (\alpha_{max} - \alpha_{min}) \tag{5.16}$$

where α_i is the step size at iteration i, $\alpha_{min}, \alpha_{max}$ are the limits for the α, and n is a constant defined according to the problem dimension as in equation 5.17.

$$n = 10^{-d} \tag{5.17}$$

where d is the problem dimension.

2. *Directed movement:* Fireflies that are not attracted by any other firefly are moved toward the best firefly according to the standard movement equation of fireflies; see equation 5.1. This is caused if there was no local best in each firefly's neighborhood; they move toward best solution and make a better position for each firefly for next iteration and they will get nearer to global best.

3. *Social behavior:* A random walk is applied on each firefly according to equation 5.18.

$$x_i = x_i + \alpha(1 - p) * U(0, 1) \tag{5.18}$$

where x_i is the firefly position, $U(0, 1)$ is a random number drawn from uniform distribution in the range $[0, 1]$, and p is calculated as

$$p = \frac{1}{\sigma\sqrt{2\pi}} e^{-\frac{(x-\mu)^2}{2\sigma^2}} \tag{5.19}$$

where μ, σ are set to 0 and 1 in order, and $x = f(best) - f(x_i)$.

5.2.9 Network-structured firefly algorithm (NS-A)

In NS-FA fireflies have a network structure [15]. The fireflies of the NS-FA are attracted to only directly connected fireflies. Therefore, the firefly of the NS-FA is not affected by a brighter firefly if there is no connection between the two fireflies. The network structure is modified with generation. The firefly of the NSFA is stochastically connected with brighter fireflies and is disconnected from less brighter fireflies. An initial network structure of the NS-FA is assumed to be ring topology. The connection among the fireflies is denoted by a connection matrix C. If the firefly i is connected to the firefly j, $C_{i,j} = 1$, otherwise $C_{i,j} = 0$.

The firefly updating equation is modified as follows:

$$x_i^{t+1} = \begin{cases} x_i^t + \beta(x_j^t - x_i^t) + \alpha(rand - 0.5)L, & C_{i,j} = 1 \text{ and } I_i > I_j \\ x_i^t, & otherwise \end{cases} \tag{5.20}$$

where L is the average scale of the problem and the other parameters are as in the standard FFA.

The firefly i, except the brightest firefly k, is stochastically disconnected from less brighter fireflies. If $Rand_i \leq C_p^t$, i is disconnected from j according to

$$C_{i,j} = 0, I_i \leq I_j, i \neq k \tag{5.21}$$

where $Rand_i$ is a random number generator uniformly distributed in [0, 1] and C_p^t is a connection probability determined by

$$C_p^t = \frac{t}{t_{max}} \tag{5.22}$$

where t, t_{max} are the iteration number and the maximum number of iterations. The network structure keeps on changing, and the fireflies converge to the optimum solution.

5.2.10 FFA with adaptive parameters

The value of γ in FFA determines the variation of attractiveness with increasing distance from communicated firefly. Using $\gamma = 0$ corresponds to no variation or constant attractiveness and conversely setting $\gamma \to \infty$ results in attractiveness being close to zero, which again is equivalent to the complete random search [16]. As mentioned in [16] *gamma* should be in the range [0, 10] and specifically γ can be estimated as

$$\gamma = \frac{\gamma_0}{r_{max}} \tag{5.23}$$

where $\gamma_0 \in [0, 1]$ and r_{max} is the maximum range in all problem dimensions.

The random term in the standard FFA updating is replaced by the following [16]:

$$u_{i,k} = \begin{cases} \alpha rand_1(b_k - x_{i,k}) & if\, sgn(rand1 - 0.5) < 0 \\ -\alpha rand_2(x_{i,k} - a_k) & otherwise \end{cases} \tag{5.24}$$

where $rand_1, rand_2$ are two random numbers from uniform distribution in the range from 0 to 1, and a_k, b_k are the limits of dimension k. The suggested alternative approach is set as a fraction of the distance to search space boundaries.

A study in [17] is performed on the parameter selection for the FFA based on experimental results on five different test functions. The author concluded that increasing the number of fireflies requires more iteration to converge and he suggests using a low number of fireflies. Best values for α and γ are around 0.1 and 0.01 in order.

A comparative study was conducted in [18] to compare the performance of FFA and particle swarm optimization (PSO) on noisy nonlinear test functions. The study finds out that the FFA is superior to PSO in reaching the global optimum. A modified version of the FFA main updating equation is proposed in [32] where the updating equation is modified as follows:

$$x_i = x_i + \beta_0 e^{-\gamma r_{ij}^2}(x_j - x_i) + \beta_0 e^{-\gamma r_{i,best}^2}(x_{best} - x_i) + \alpha(rand - 0.5) \qquad (5.25)$$

where the second term is due to the attraction while the fourth term is randomization with α being the randomization parameter. $rand$ is a random number generator uniformly distributed in $[0, 1]$. β_0 is always set to 1 $\alpha \in [0, 1]$. Furthermore, the randomization term can easily be extended to a normal distribution $N(0, 1)$ and x_{best} is the position of best solution.

5.3 FFA hybridizations

In literature many hybridizations are performed between firefly algorithm and other optimizers to enhance its performance in solving optimization tasks. Some of the motivations as mentioned in the literature are mentioned below.

- To improve the population diversity and to avoid the premature convergence of FFA.

- To minimize the number of function evaluations is a main problem in FFA.

- To avoid falling in local minima.

Table 5.1 states samples of hybridizations mentioned in the literature.

5.3.1 Hybrid evolutionary firefly algorithm (HEFA)

The HEFA is a hybrid between FFA and differential evolution (DE) and is proposed in [20]. In this proposal the population in each iteration is divided into two groups and each group is updated individually as follows:

Table 5.1: FFA hybridizations and sample references

Hybrid with	Target	Sample Reference
Evolutionary algorithm (EA)	Enhance global searching	[20]
Genetic algorithms (GA)	Enhance global searching	[21–23]
Harmony search (HS)	Explore the new search space and exploit the population	[24]
Pattern search (PS)	Make use of local searching capability of PS	[25]
Learning automata (LA)	Adapt FFA parameters	[26]
Ant colony	Avoid falling in local minima	[22]

- *Agents with potential fitness:* This group is handled as in the standard firefly algorithm where each firefly is moved to the brighter fireflies.

- *Agents with less significant fitness:* The solutions in this population undergo the evolutionary operations of the DE method. Mutation operation is performed on the original solution generating a trivial solution v_i according to the following equation:

$$v = x_{best} + F(x_{r1} - x_{r2}) \tag{5.26}$$

where x_{best} is the best solution, F is a mutation constant, and x_{r1}, x_{r2} are two random solutions from the group.

Offspring solution is produced by the crossover operation that involved the parent x_i and the trivial solution v_i. The vectors of the ith offspring solution, yi, are created as follows:

$$y_i = \begin{cases} v_i & if R < CR \\ x_i & otherwise \end{cases} \tag{5.27}$$

where R is a random number drawn from uniform distribution $\in [0,1]$ and CR is a constant crossover coefficient.

Each new solution is compared to its original one and the better one is kept as in equation 5.28.

$$x_i^{t+1} = \begin{cases} x_i^t & if f(x_i^t) < f(y_i) \\ y_i & otherwise \end{cases} \tag{5.28}$$

This indicates that the original solution would be replaced by the offspring solution if the fitness value of the offspring solution were better than the original solution. Otherwise, the original solution would remain in the population for the next iteration.

Another variant of FFA is proposed in [21] that is more similar to HEFA. The global searching power of a genetic algorithm is employed in the standard FFA which is much better in a local search. In this proposal all solutions in each iteration are used to generate new ones based on the standard GA steps, namely, crossover, mutation, and selection.

Another variant of hybridization between FFA and genetic is proposed in [22]. In each iteration of the genetic optimization the population generated is passed to FFA updating. Genetic algorithm searches the solution space for global minimum and the firefly algorithm improves the precession of the potential candidate solution.

A hybrid algorithm between FFA and GA is proposed in [23]. Each individual optimizer is initialized independently and at each iteration the population of each algorithm is swapped to take advantage of both the global search capability of GA and the local searching capability of FFA.

5.3.2 Hybrid FFA and harmony search (HS/FA)

A hybrid by inducing harmony search (HS) into the FA method named HS/FA is utilized to deal with the optimization problem, which can be considered as mutation operator [24]. The mutation of the HS and FA can explore the new search space and exploit the population, respectively. Therefore, it can overcome the lack of exploration of the FA [24].

The hybrid algorithm performs attraction only to a number of best fireflies, not all of which speeds up the number of function evaluations. The second improvement is the addition of HS serving as mutation operator striving to improve the population diversity to avoid the premature convergence. In HS/FA, if firefly i is not brighter than firefly j, firefly j is updated by mutation operation to improve the light intensity for firefly j. Every element $x_{kj}(k = 1, 2, ..., D)$ is tuned as follows:

$$x_{\nu,k} = \begin{cases} x_{\nu,k} + bw(2rand - 1) & \text{if } rand_1 < HMCR \text{ and } rand_2 < PAR \\ x_{r1,k} & \text{if } rand_1 < HMCR \text{ and } \geq PAR \\ x_{min,k} + rand(x_{max,k} - x_{min,k}) & otherwise \end{cases}$$

(5.29)

- x_{r1} is a random solution drawn from the current fireflies randomly.

- $rand$ is uniform random number in [0,1].

- bw is band width.

- $HMCR$ is the harmony memory consideration rate.

- PAR is the pitch adjustment rate.

- x_{min}, x_{max} are the limits for the search space.

5.3.3 Hybrid FFA and pattern search (hFAPS)

In [25] a hybrid FFA and pattern search optimizer is used for automatic generation control (AGC) of multi-area power systems with the consideration of generation rate constraint (GRC). FFA is used to find the parameters of proportional integral derivative (PID) controllers employing an integral time multiple absolute error (ITAE) fitness function. Pattern search (PS) for its local searching power is then employed to fine tune the best solution provided by FA.

5.3.4 Learning automata firefly algorithm (LA-FF)

In firefly algorithm, for any two flashing fireflies, the less bright one will move toward the brighter, with a percentage of randomness. The initial percentage value and other parameter such as attractiveness coefficient are crucially important in determining the speed of the convergence and how the FA algorithm behaves. α determines random percentage in firefly moving. Attractiveness coefficient is named γ. The parameter varies between zero to extreme [26]. Absorption coefficient is set by one learning automata. In each iteration, the value of parameter is set as

- Unchanged

- Increased by multiplying a number greater than one

- Increased by multiplying a number smaller than one

In any iteration, the learning automaton functions three actions: decrease, increase, and unchanged gamma parameter; therefore, any action determines the parameter value related to this value in the previous iteration. In the learning automata's problem, environment with input α changes probability of any action according to a set of rules. Learning automata environment considers P model and it is applied from learning algorithm $p(n+1) = T[\alpha(n), \beta(n), p(n)]$ [26].

5.3.5 Hybridization between ant colony optimization and FFA

To avoid falling in a local minima, ant colony optimization (ACO) is enhanced using the local searching power of FFA [22]. The pheromone generated by the ACO is converted into candidate solutions using equation 5.30.

$$p_{ij}^k(t) = \begin{cases} \frac{[\tau_{ij}(t)]^\alpha}{\sum_{l \in allowed_k}[\tau_{il}(t)]^\alpha} & \text{for all } x_{ij} \text{ , } x_{ij} \in allowed_k \\ 0, & otherwise \end{cases} \tag{5.30}$$

where $x_{ij}^k(t)$ is probability that option x_{ij} is chosen by ant k for variable i at time t and $allowed_k$ is the set of the candidate values contained in the group i.

The candidate ACO solutions are used to initialize the FFA for further enhancement. The random component in the standard FFA updating equation is adapted as follows:

$$\alpha_{t+1} = \alpha_t \theta^{1-\frac{t}{Tc}} \tag{5.31}$$

where t is iteration number, Tc maximum number of iterations, and $\theta \in [0,1]$ is the randomness reduction constant.

5.4 Firefly in real world applications

This section presents sample applications for FFA. FFA is successfully applied in many disciplines in decision support and decision making, in the engineering field, computer science, and communication. The main focus is on identifying the fitness function and the optimization variables used in individual applications. Table 5.2 summarizes sample applications and their corresponding objective(s).

In [28] FFA is used to tackle the problem of *queuing*. Queueing theory provides methods for analysis of complex service systems in computer systems, communications,

Table 5.2: FFA applications and the corresponding fitness function

Application Name	Fitness Function	Sample References
Queuing problem	Maximizes the overall profits	[28]
Job scheduling	Minimize the make span time as well as the flow time	[21]
Quadratic assignment problem (QAP)	minimum total cost	[29]
Data clustering	Maximum entropy	[30–33]
Data clustering	Minimize fuzzy c-means fitness function	[35]
Cloud computing	Maximize the resource utilization and provide a good balanced load among all the resources in cloud servers	[36]
Image retrieval	Feature weighting to maximize correlation between query and content	[37]
Proportional-integral-derivative (PID)	Goal is to improve the automatic voltage regulator system step response characteristics	[38]
System identification	Objective is to find a model and a set of parameters that minimize the prediction error between system output $y(t)$, measured data, and model output $y(\hat{t}, \theta)$ at each time step t	[34, 39]
Economic dispatch (ED)	Minimize the *total system cost* where the total system cost is a function composed by the sum of the cost functions of each online generator	[40–42]
UCAV path planning	Minimize threat and fuel consumption	[43]
Structural mass optimization	Frequency constraint weight minimization	[44, 45]
Traveling salesman problem (TSP)	Minimize length of the tour	[46]
Social portfolio problem	Make a proper selection of projects to benefit from the portfolio of projects to optimize properly, the social benefit for each of the minorities with the aim of improving their living conditions in everyday aspects	[47]
Supplier selection problem	Target is the select supplier(s) to provide a given quantity with minimum transportation cost and product price	[48]
Mathematical programming applications	Create multiple solution alternatives that both satisfy required system performance criteria and yet are maximally different in their decision spaces	[49]
Cluster head selection	Main issue is power consumption and lifetime of wireless sensor network	[50]

transportation networks, and manufacturing. In this problem the goal is to seek the number of servers that maximizes the overall profits.

The *job scheduling* is an NP-hard problem that is solved in [5] using FFA. The objective is to minimize the *make span time* as well as *the flow time*. The job scheduling targets to distribute optimally a set of N tasks on M resources. The same problem is handled in [21] but using a hybrid algorithm between FFA and genetic algorithm to make use of the global searching power of the genetic algorithm as well as to make use of the local searching capability of the FFA.

The *quadratic assignment problem* (QAP) is a mathematical model for the location of indivisible economic activities, and is one of the basic computational problems in computer science with complexity strong NP-hard [29]. For the given n facilities and n locations, with the given flows between all the pairs of the facilities and the given distances between all the pairs of the locations, find an assignment of each facility to the unique location, such that the *total cost*, computed as the sum of distances times corresponding flows, is minimum. The problem is faced in [29] using FFA. The problem is formulated as follows:

$$\min \sum_{i=1}^{n} \sum_{j=1}^{n} a_{ij} b_{\pi(i)\pi(j)} \tag{5.32}$$

where

- a_{ij} represents the flow from the facility i to the facility j.

- b_{ij} represents the distance from the location i to the location j.

- n is the number of facilities or locations.

A discrete version of FFA is used to optimize the QAP in [29].

In [30,31] FFA is used to create initial cluster centers for the K-means clustering. The used fitness for FFA is maximum entropy as in equation 5.33.

$$f(x) = \frac{1}{\sum_{i=1}^{k} S_i} \tag{5.33}$$

where k is the number of clusters and S_i is defined as

$$S_i = \sum_{j \in C_i} [-\frac{P_{ij}}{P_i}.log(\frac{P_{ij}}{P_j})] \tag{5.34}$$

where P_{ij} is the probability of occurrence of pixel value j in cluster i and P_i is the probability of occurrence of all the pixel values in class i. This used initialization overcomes most of the problems related to K-means, especially speed, clustering accuracy, and local minima problems [30,32]. In [33] FFA with K-means is used for Dorsal Hand Vein segmentation from the CCD camera.

The fuzzy c-means (FCM) fitness function is used with FFA for image data clustering in [35] to overcome the local minima problem in the standard FCM.

Vein structure is unique to every individual. There are different imaging methods where near-infrared lighting sensors are more common in capturing dorsal hand vein, palm vein, and fingers vein patterns. Black and white CCD cameras are also sensitive in the near-infrared region, so a filter blocking the visible light is all that is needed on the camera. Since shape and state of skin have no effect on the system's result, dorsal hand vein patterns are more secure than finger print and hand geometry [33]. These are one of the highest popular systems among security and biometric systems because of their uniqueness and stability [33].

Image registration is the process of transforming different images into the same co-ordinate system [8]. In [8] an image registration model is proposed based on FFA. A chaotic FFA is used to find the optimal transformation parameter set that maximizes *the normalized cross correlation (NCC)*.

$$\min_{[Tx,Ty,\theta]} \frac{1}{N} \sum_{x,y} \frac{(A - \mu_A)(T(B) - \mu_{T(B)})}{\sigma_A \sigma_{T(B)}} \tag{5.35}$$

where N denotes the total number of pixels in image A and B, μ_A and μ_B denote the average of the images A, B, and σ_A, σ_{TB} denote the standard deviation for image A and transformed image B.

A *cloud computing* is conceptually a distributed and elastic system where resources will be distributed through the network. The full resources of the system cooperate to respond to a client request which requires intercommunication between various components of the system to design a component or subset of components to deal with the request [36]. This can lead to bottlenecks in the network and an imbalanced charge in a distributed system where some components will be overcharged while others will not or will be light charged. Among the major challenges faced by the cloud computing systems, *load balancing* is one tedious challenge that plays a very important role in the realization of cloud computing. For developing strategy for load balancing, the main points to be considered are estimation of load, comparison of load, stability of different system, performance of system, interaction between the nodes, nature of work to be transferred, and selecting of nodes. Thus, in order to tackle the problems related to load balancing in the cloud network, an effective system should be developed. The design of a load balancing model using firefly algorithm is proposed in [36]. The main objective is to maximize the *resource utilization* and provide a *good balanced load* among all the resources in cloud servers.

In [37] FFA is used in the content-based image retrieval (CBIR). The CBIR system has been implemented with the relevance feedback mechanism and for the optimization the objective function for the firefly has been designed with image color parameters and texture parameters. The decision variables of the firefly algorithm are the four feature vectors $M_{ch}, M_{cm}, M_{edh}, M_{wt}$. The brightness of light intensity is associated with the objective function, which is related to the sum of weighted Euclidean distance between the query image and the stored database image in D-dimensional search space: M_{ch} color histogram bins, M_{cm} color moments, M_{edh} edge direction histogram, and M_{wt} wavelet texture feature values.

The work in [38] presents a tuning approach based on continuous firefly algorithm (CFA) to obtain the *proportional-integral-derivative (PID)* controller parameters in an automatic voltage regulator system (AVR). In the tuning processes the CFA is iterated to reach the optimal or the near-optimal of PID controller parameters when the main goal is to improve the AVR step response characteristics. In time domain, the fitness can be formed by different performance specifications such as the *integral of time multiplied by absolute error value (ITAE), rise time, settling time, overshoot,* and *steady-state error.*

The problem of determining a mathematical model for an unknown system by monitoring its input-output data is known as *system identification.* This mathematical model then could be used for control system design purposes [34]. A modified version of FFA is used in [34] for the system identification problem.

System identification consists of two subtasks:

- Structural identification of the equations in the model M

- Parameter identification of the model's parameter $\hat{\theta}$

A system identification problem can be formulated as an optimization task where the objective is to find a model and a set of parameters that minimize the prediction error between system output $y(t)$, measured data, and model output $y(\hat{t}, \theta)$ at each time step t; see equation 5.36 [39].

$$MSE(\hat{y}, \hat{\theta}) = \frac{1}{T} \sum_{t=1}^{T} (y(t) - \hat{y}(t, \hat{\theta}))^2 \tag{5.36}$$

In [40–42] FFA is used to handle the *economic dispatch (ED)* problem. The ED problem is an optimization problem that determines the power output of each online generator that will result in a least cost system operating state. The objective of economic dispatch is to minimize the *total system cost* where the total system cost is a function composed of the sum of the cost functions of each online generator. This power allocation is done considering system balance between generation and loads, and feasible regions of operation for each generating unit. The objective of the classical economic dispatch is to minimize the total fuel cost by adjusting the power output of each of the generators connected to the grid.

FFA is used in [43] for *path planning* for *uninhabited combat air vehicles* (UCAVs). UCAV is a complicated high dimension optimization problem, which mainly centralizes on optimizing the flight route considering the different kinds of constraints under complicated battlefield environments. A performance indicator of path planning for UCAV mainly contains the *completion of the mandate of the safety performance indicator* and *fuel performance indicator*, i.e., indicators with the least threat and the least fuel [43].

Structural mass optimization on shape and size is performed in [44] using FFA taking into account dynamic constraints. In this kind of problem, *mass reduction* especially conflicts with frequency constraints when they are lower bounded, since vibration modes may easily switch due to shape modifications. The main goal is that engineering structures are often supposed to be as light as possible. Thus a frequency constraint weight optimization process should be performed to obtain these two aims simultaneously.

In design optimization of truss structures, the objective is to find the best feasible structure with a minimum weight [45]. Optimum design of truss structures is a search for the best possible arrangements of design variables according to the determined constraints. Design variables involved in optimum design of truss structures can be considered as sizing, geometry, and topology variables. In sizing optimization of truss structures, the aim is to find the optimum values for cross-sectional areas of the elements. Geometry optimization means to determine the optimum positions of the nodes while presence or absence of the members are considered in the topology optimization. FFA is used in [45] for the optimization of truss structures.

The goal of the traveling salesman problem (TSP) is to find a tour of a given number of cities, visiting each city exactly once and returning to the starting city where the *length of the tour* is minimized. The TSP is a NP-hard problem [46]. A discrete version of FFA is used to solve the TSP problem in [46].

In [47] FFA is used in decision making about a social portfolio known in English as "*Social Portfolio Problem*" based on a social modeling characterized by four minorities. In this social representation it is necessary to make a proper selection of projects to benefit from the portfolio of projects to optimize properly, the social benefit for each of the minorities with the aim of improving their living conditions in everyday aspects. The fitness function is formulated in the following equation:

$$K = IS[(PE - IE)^{PCM}] \pm RSLP \tag{5.37}$$

where IS = social impact of the project; PE = economic potential of the project; IE = ecological impact; PCM = minority cultural preservation, since the project does not affect the culture associated with the minority; and $RSLP$ = social return of long-term project, meaning that the project changed over time the situation of the minority.

A discrete version of FFA is used in [48] for a *supplier selection problem*. The supplier selection is a crucial problem of the logistics. The required quantity of the product is given: Q. The minimum and maximum quantities are also given for every supplier. The capacity of the transport vehicles are P_i, which is constant in the model for every transport vehicle at the given supplier. The transportation cost is also defined for every vehicle: Tr_i. The target is select supplier(s) to *provide a given quantity with minimum transportation cost and product price* according to the equation

$$minC = \sum_{i=1}^{n}(C_i^T + C_i^{TR}) \tag{5.38}$$

where C_i is purchasing price of the product and C_i^{TR} is transportation cost.

When solving many practical *mathematical programming applications*, it is generally preferable to formulate numerous quantifiably good alternatives that provide very different perspectives to the problem [49]. These alternatives should possess near-optimal objective measures with respect to all known modeled objectives, but be fundamentally different from each other in terms of the system structures characterized by their decision variables. This solution approach is referred to as modeling to generate alternatives (MGA). The Firefly algorithm was used to efficiently create multiple solution alternatives that both satisfy required *system performance* criteria and yet are maximally *different in their decision spaces* in [49].

A *cluster head selection* for a wireless sensor network (WSN) is proposed in [50]. In wireless sensor network the main issue is power consumption and lifetime of network. This can be achieved by selection of proper cluster heads in a cluster-based protocol. The selection of cluster heads and its members is an essential process which affects energy consumption. A hybrid clustering approach proposes to minimize the energy of the network so the lifetime of WSN can be increased.

5.4.1 Firefly algorithm in feature selection

FFA was used to select optimal feature combination to maximize classification performance in this case study. Eight datasets from the UCI machine learning repository [51] are used in the experiments and comparison results. The eight datasets were selected to have various numbers of attributes and instances as representatives of various kinds of issues that the proposed technique would be tested on, as shown in Table 5.3.

For each dataset, the instances are randomly divided into three sets, namely, *training*, *validation*, and *testing* sets. The training set is used to train the used classifier, while the validation set is used to evaluate the classifier performance and is used inside the optimization fitness. The test data are kept hidden for both the classifier and the optimizer for the final evaluation of the whole feature selection and classification system.

Table 5.3: Datasets used in the case study

Dataset	No. of Attributes	No. of Instances
Breast cancer	9	699
Exactly	13	1000
Exactly2	13	1000
Lymphography	18	148
M-of-N	13	1000
Tic-tac-toe	9	958
Vote	16	300
Zoo	16	101

The fitness function for the FFA is to maximize classification performance over the validation set given the training data 5.39.

$$\downarrow_x \frac{1}{N} \sum_{i=1}^{N} |O_i - R_i| \qquad (5.39)$$

where x is the current selected features, N is the number of data points in the validation set, and O_i, R_i are the output classifier result and the reference true classification result. Individual solutions in the FFA are points in the feature space; d-dimensional space, where d is the number of features in the original dataset in the range $[0, 1]$. The well-known K-nearest neighbor classifier was used in the fitness function. The FFA is randomly initialized with solutions in the feature space and is applied to minimize the fitness function in equation 5.39 and given the following parameter set. It worth mentioning that the randomness parameter α is decremented by a factor δ at each iteration as in the following equation:

$$\alpha^t = \delta * \alpha^{t-1} \qquad (5.40)$$

The genetic algorithm (GA) [52] optimizer and particle swarm optimizer (PSO) [53] are used in the same manner to be compared with the FFA to evaluate its classification performance.

Table 5.4: FFA parameter setting

Parameter	Value	Meaning
α	0.5	Randomness parameter
γ	1	Absorbtion coefficient
δ	0.99	Randomness reduction coefficient
$\beta 0$	0.2	Light amplitude

Table 5.5: Best, mean, and worst obtained fitness value for the different optimizers used

Dataset	GA			PSO			FPA		
	Best	*Mean*	*Worst*	*Best*	*Mean*	*Worst*	*Best*	*Mean*	*Worst*
Breast cancer	0.0215	0.0240	0.0258	0.0129	0.0223	0.0258	0.0129	0.0232	0.0300
Exactly	0.2485	0.2844	0.3054	0.2784	0.2892	0.3024	0.0539	0.1485	0.2814
Exactly2	0.2186	0.2293	0.2425	0.2096	0.2281	0.2455	0.2096	0.2257	0.2515
Lymphography	0.1224	0.1557	0.2041	0.1224	0.1598	0.2244	0.1224	0.1352	0.1837
M-of-N	0	0.0766	0.1198	0	0.0808	0.1886	0	0.0108	0.0299
Tic-tac-toe	0.1781	0.2238	0.2531	0.1781	0.23625	0.2656	0.2062	0.225	0.2531
Vote	0.02	0.042	0.06	0.03	0.044	0.07	0.02	0.032	0.05
Zoo	0	0.0531	0.0882	0	0.0427	0.0882	0	0.0357	0.1176

Table 5.5 displays the best, mean, and worst fitness function value obtained by running each stochastic algorithm for 10 different initializations. We can see that the FFA obtains much enhanced fitness value over both PSO and GA. The advance in the obtained fitness value is in the best, mean, and worst obtained solution.

The performance of the different optimizers over the different test dataset is outlined in Figure 5.1. We can remark from the figure that the performance of FFA is much better than PSO and GA, which proves that the selected feature combinations are much better.

Table 5.6 represents the average number of features selected to the total number of features for different datasets and using different optimizers. We can remark that FFA selects a minimum number of features in comparison with PSO and GA, while it keeps better classification performance as outline in Table 5.5.

For assessing the stability of the performance of the optimizer we run an individual optimizer for 10 different running and using different initial random solutions and measuring the standard deviation of the obtained best solution. Table 6.5 represents the calculated standard deviation of the obtained fitness function. We can remark that the obtained variation for the FFA is comparable to both GA and PSO, although the random factor for FFA is set to 0.5, a respectively high value. We allowed for high randomization factor or random factor in the FFA to tolerate for the high variation existing in the search space so that FFA can reach the global minima.

Table 5.6: Average feature reduction for different data sets using different optimizers

Dataset	*bcancer*	*exactly*	*exactly2*	*Lymph*	*m-of-n*	*TicTacToe*	*Vote*	*Zoo*
FFA	0.6	0.4462	0.3231	0.3333	0.4615	0.6222	0.2875	0.375
GA	0.7111	0.4769	0.5077	0.4556	0.6462	0.6667	0.4375	0.6125
PSO	0.6444	0.7846	0.3385	0.5444	0.5692	0.6444	0.475	0.55

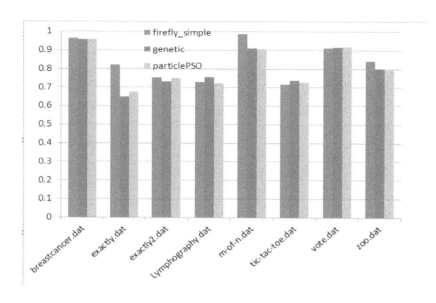

Figure 5.1: Classification accuracy over the different data sets and using different optimizers

Table 5.7: Obtained standard deviation for the different datasets and using different optimizers

Dataset	*bcancer*	*exactly*	*exactly2*	*Lymph*	*m-of-n*	*TicTacToe*	*Vote*	*Zoo*
GA	0.0024	0.0222	0.0091	0.0307	0.0473	0.0290	0.0148	0.03821
PSO	0.0056	0.0109	0.0135	0.039	0.0730	0.0354	0.0152	0.0405
FFA	0.0078	0.1202	0.0171	0.0271	0.0149	0.0173	0.0164	0.0483

5.5 Chapter conclusion

This chapter presented the main behavior of fireflies, the artificial FFA, and its variants that were proposed in 2008. Real applications of FFA were reviewed and the chapter ended by showing results of the algorithm in a featured selection problem.

Bibliography

[1] X.S. Yang, "Firefly algorithm". *Nature-Inspired Metaheuristic Algorithms*, Luniver Press, UK, 2008.

[2] X.-S. Yang, "Firefly algorithms for multi-modal optimization", in: *Stochastic Algorithms: Foundations and Applications*, SAGA 2009, Lecture Notes in Computer Sciences, Vol. 5792, pp. 169-178, 2009.

[3] Adil Hashmi, Nishant Goel, Shruti Goel, and Divya Gupta, "Firefly algorithm for unconstrained optimization", *IOSR Journal of Computer Engineering* (IOSR-JCE), Vol. 11(1), pp. 75-78, 2013.

[4] Karel Durkota, "Implementation of Discrete Firefly Algorithm for the QAP Problem within the Seage Framework", Czeck Technical University in Prague, Faculty of Engineering, 2011.

[5] Adil Yousif, Abdul Hanan Abdullah, Sulaiman Mohd Nor,Adil Ali Abdelaziz, "Scheduling Jobs On Grid Computing Using Firefly Algorithm", *Journal of Theoretical and Applied Information Technology*, vol 33, No 2, pp 155-164.

[6] R. Falcon, M. Almeida, and A. Nayak, "Fault identification with binary adaptive fireflies in parallel and distributed systems", in: 2011 IEEE Congress on Evolutionary Computation (CEC), Vol. 11, pp. 1359-1366, 2011.

[7] K. Chandrasekaran and S. Simon, "Network and reliability constrained unit commitment problem using binary real coded firefly algorithm", *International Journal of Electrical Power and Energy Systems*, Vol. 43(1), pp. 921-932, 2012.

[8] Yudong Zhang and Lenan Wu, "Rigid image registration based on normalized cross correlation and chaotic firefly algorithm", *International Journal of Digital Content Technology and Its Applications* (JDCTA), Vol. 6(22), pp. 129-138, December 2012.

[9] L. dos Santos Coelho, D. L. de Andrade Bernert, and V. C. Mariani, "A chaotic firefly algorithm applied to reliability-redundancy optimization", in: 2011 IEEE Congress on Evolutionary Computation (CEC), pp. 517-521, 2011.

[10] M. Subutic, M. Tuba, and N. Stanarevic, "Parallelization of the firefly algorithm for unconstrained optimization problems", in: *Latest Advances in Information Science and Applications*, pp. 264-269, 2012.

[11] A.M. Deshpande, G.M. Phatnani, and A.J. Kulkami, "Constraint handling in firefly algorithm", in: 2013 IEEE International Conference on Cybernetics (CYBCONF), 13-15 June 2013, Lausanne, pp. 186-190, 2013.

[12] X. S. Yang, "Firefly algorithm, levy flights and global optimization", in: *Research and Development in Intelligent Systems XXVI*, Springer, pp. 209-218, 2010.

[13] Seif-Eddeen K. Fateen and Adrin Bonilla-Petriciolet, "Intelligent firefly algorithm for global optimization, cuckoo search and firefly algorithm studies", *Computational Intelligence*, Vol. 516, pp. 315-330, 2014.

[14] Sh. M. Farahani, A. A. Abshouri, B. Nasiri, and M. R. Meybodi, "A gaussian firefly algorithm", *International Journal of Machine Learning and Computing*, Vol. 1(5), pp. 448-453, December 2011.

[15] Daiki Matsumoto and Haruna Matsushita, "Network-Structured Firefly Algorithm", in: IEEE Workshop on Nonlinear Circuit Networks December 14-15, pp. 48-50, 2012.

[16] S. Lukasik and S. Zak, "Firefly algorithm for continuous constrained optimization tasks, in: Proceedings of the International Conference on Computer and Computational Intelligence (ICCCI09)", N. T. Nguyen, R. Kowalczyk, and S.-M. Chen, Eds., Vol. 5796 of LNAI, pp. 97-106, Springer, Wroclaw, Poland, October 2009.

[17] Sankalap Arora and Satvir Singh, "The firefly optimization algorithm: convergence analysis and parameter selection", *International Journal of Computer Applications* (0975-8887), Vol. 69(3), pp. 48-52, May 2013.

[18] Saibal K. Pal, C.S. Rai, and Amrit Pal Singh, "Comparative study of firefly algorithm and particle swarm optimization for noisy non-linear optimization problems", *International Journal Intelligent Systems and Applications*, Vol. 10, pp. 50-57, 2012.

[19] T. Hassanzadeh, Iran Qazvin, and M.R. Meybodi, A new hybrid approach for data clustering using firefly algorithm and K-means, International Symposium on Artificial Intelligence and Signal Processing (AISP), Shiraz, Fars, pp. 7-11, 2012.

[20] Afnizanfaizal Abdullah, Safaai Deris, Mohd Saberi Mohamad, and Siti Zaiton Mohd Hashim, A New Hybrid Firefly Algorithm for Complex and Nonlinear Problem, Distributed Computing and Artificial Intelligence, AISC 151, Springer-Verlag. Berlin Heidelberg Germany, pp. 673-680, 2012.

[21] Hossein Parvan, Ebrahim Behrouzian Nejad, and Seyed Enayatolah Alavi, "New hybrid algorithms for task scheduling in computational grids to decrease makespan", *International Journal of Computer Science and Network Solutions*, Vol. 2(4), pp. 19-28, April 2014.

[22] Shaik Farook and P. Sangameswara Raju, "Evolutionary hybrid genetic-firefly algorithm for global optimization", *IJCEM International Journal of Computational Engineering and Management*, Vol. 16(3), May 2013.

[23] Sh. M. Farahani, A. A. Abshouri, B. Nasiri, and M. R. Meybodi, "Some hybrid models to improve firefly algorithm performance", *International Journal of Artificial Intellegence*, Vol. 8(s12), 2012.

[24] Lihong Guo, Gai-Ge Wang, Heqi Wang, and Dinan Wang, "An effective hybrid firefly algorithm with harmony search for global numerical optimization", *The Scientific World Journal*, Vol. 2013, Article ID 125625, pp. 1-9, http://www.hindawi.com/journals/tswj/2013/125625/.

[25] Rabindra Kumar Sahu, Sidhartha Panda, and Saroj Padhan, "A hybrid firefly algorithm and pattern search technique for automatic generation control of multi area power systems", *International Journal of Electrical Power and Energy Systems*, Vol. 64, pp. 9-23, January 2015.

[26] S. M. Farahani, A. A. Abshouri, B. Nasiri, and M. R. Meybodi, "Some hybrid models to improve firefly algorithm performance", *International Journal of Artificial Intelligence*, Vol. 8, pp. 97-117, 2012.

[27] Ahmed Ahmed El-Sawy, Elsayed M. Zaki, and R. M. Rizk-Allah, "Hybridizing Ant colony Optimization With Firefly Algorithm For Unconstrained Optimization Problems", *The Online Journal on Computer Science and Information Technology* (OJCSIT), Vol.(33) pp. 185-193, 2013.

[28] J. Kwiecie and B. Filipowicz, "Firefly algorithm in optimization of queueing systems", *Bulletin of the Polish Academy of Sciences, Technical Sciences*, Vol. 60(2), pp. 363-368, 2012.

[29] Karel Durkota, "Implementation of a discreet firefly algorithm for the QAP problem within the SEAGE framework", Czeck Technical University in Prague, Faculty of Electrical Engineering 2011.

[30] Yang Jie, Yang Yang, Yu Weiyu, and Feng Jiuchao, "Multi-threshold Image Segmentation Based on K-means and Firefly Algorithm", in: Proceedings of the 3rd International Conference on Multimedia Technology, ICMT13, pp. 134-142, 2013.

[31] Bhavana Vishwakarma and Amit Yerpude, "A new method for noisy image segmentation using firefly algorithm", *International Journal of Science and Research*, Vol. 3(5), pp. 1721-1725, May 2014.

[32] T. Hassanzadeh and M.R. Meybodi, "A new hybrid approach for data clustering using firefly algorithm and K-means", International Symposium on Artificial Intelligence and Signal Processing (AISP), Shiraz, Fars, pp. 7-11, 2012.

[33] Zahra Honarpisheh and Karim Faez, "An efficient dorsal hand vein recognition based on firefly algorithm", *International Journal of Electrical and Computer Engineering* (IJECE), Vol. 3(1), pp. 30-41, February 2013.

[34] Mehrnoosh Shafaati and Hamed Mojallali, "Modified firefly optimization for IIR system identification", *CEAI*, Vol. 14(4), pp. 59-69, 2012.

[35] Parisut Jitpakdee, Pakinee Aimmanee, and Bunyarit Uyyanonvara, "Image Clustering Using Fuzzy-based Firefly Algorithm", in: 5th International Conference on Computer Research and Development, (ICCRD 2013) Ho Chi Minh City, Vietnam, 23-24 February 2013.

[36] A. Paulin Florence and V. Shanthi, "A load balancing model using firefly algorithm in cloud computing", *Journal of Computer Science*, Vol. 10(7), 2014.

[37] T. Kanimozhi and K. Latha, "A Meta-Heuristic Optimization Approach for Content Based Image Retrieval Using Relevance Feedback Method", in: Proceedings of the World Congress on Engineering, Vol. (II), July 3-5, London, U.K., 2013.

[38] Omar Bendjeghaba, "Continuous firefly algorithm for optimal tuning of PID controller in AVR system", *Journal of Electrical Engineering* Vol. 65(1), pp. 44-49, 2014.

[39] Rasmus K. Ursem, "Models for Evolutionary Algorithms and Their Applications in System Identification and Control Optimization", Department of Computer Science, University of Aarhus, Denmark, 2003.

[40] R. Subramanian and K. Thanushkodi, "An efficient firefly algorithm to solve economic dispatch problems", *International Journal of Soft Computing and Engineering* (IJSCE), Vol. 2(1), pp. 52-55, March 2013.

[41] K. Sudhakara Reddy and M. Damodar Reddy, "Economic load dispatch using firefly algorithm", *International Journal of Engineering Research and Applications* (IJERA), Vol. 2(4), pp. 2325-2330, July-August 2012.

[42] Xin-She Yang, Seyyed Soheil Sadat Hosseini, and Amir Hossein Gandomi, "Firefly algorithm for solving non-convex economic dispatch problems with valve loading effect", *Applied Soft Computing*, Vol. 12, pp. 1180-1186, 2012.

[43] Gaige Wang, Lihong Guo, Hong Duan, Luo Liu, and Heqi Wang, "A modified firefly algorithm for UCAV path planning", *International Journal of Hybrid Information Technology*, Vol. 5(3), pp. 123-144, July, 2012.

[44] Herbert M. Gomes, "A firefly metaheuristic algorithm for structural size and shape optimization with dynamic constraints", *Mecinica Computational*, Vol. 30, pp. 2059-2074, Rosario, Argentina, 1-4 November 2011.

[45] S. Kazemzadeh Azada and S. Kazemzadeh Azad, "Optimum design of structural using an improved firefly algorithm", *International Journal of Optimization in Civil Engineering*, Vol. 2, pp. 327-340, 2011.

[46] Sharad N. Kumbharana1 and Gopal M. Pandey, "Solving travelling salesman problem using firefly algorithm", *International Journal for Research in Science and Advanced Technologies*, Vol. 2(2), pp. 53-57, 2014.

[47] Alberto Ochoa-Zezzatti, Alberto Hernndez, and Julio Ponce, "Use of Firefly Algorithm to Optimize Social Projects to Minorities", in: Fourth World Congress on Nature and Biologically Inspired Computing (NaBIC), 2012.

[48] Laszlo Kota, "Optimization of the supplier selection problem using discrete firefly algorithm", *Advanced Logistic Systems*, Vol. 6(1), pp. 117-126, 2012.

[49] Raha Imanirad, Xin-She Yang, and Julian Scott Yeomans, "Modelling-to-generate-alternatives via the firefly algorithm", *Journal of Applied Operational Research*, Vol. 5(1), pp. 14-21, 2013.

[50] P. Leela and K. Yogitha, "Hybrid Approach for Energy Optimization in Wireless Sensor Networks", in: 2014 IEEE International Conference on Innovations in Engineering and Technology (ICIET14), 21-22 March, Organized by K.L.N. College of Engineering, Madurai, Tamil Nadu, India.

[51] A. Frank and A. Asuncion, UCI Machine Learning Repository, 2010.

[52] A.E. Eiben et al., "Genetic algorithms with multi-parent recombination", Proceedings of the International Conference on Evolutionary Computation (The Third Conference on Parallel Problem Solving from Nature), pp. 78-87, 1994.

[53] J. Kennedy and R. Eberhart, Proceedings of IEEE International Conference on Neural Networks, IV, pp. 1942-1948, 1995, doi:10.1109/ICNN.1995.488968.

6 Flower Pollination Algorithm

6.1 Flower pollination algorithm (FPA)

The flower search is a meta-heuristic optimization algorithm based on the flower pollination process of flowering plants that was proposed by Yang [1]. FPA uses the same principles of pollination that exist in nature.

6.1.1 Characteristics of flower pollination

The main purpose of a flower is ultimately reproduction via pollination. Flower pollination is typically associated with the transfer of pollen; such transfer is often linked with pollinators such as insects, birds, bats, and other animals [2].

Pollination can take two major forms: *abiotic* and *biotic*. About ninety percent of a flowering plants belong to biotic pollination, that is, pollen is transferred by a pollinator such as insects and animals. About ten percent of pollination takes abiotic form which does not require any pollinators. Pollination can be achieved by *self-pollination* or *cross-pollination*. *Cross-pollination*, or allogamy, means pollination can occur from pollen of a flower of different plant, while *self-pollination* is the fertilization of one flower, such as peach flowers, from pollen of the same flower or different flowers of the same plant, which often occurs when there is no reliable pollinator available.

Biotic, cross-pollination may occur at long distance, and the pollinators such as bees, bats, birds, and flies can fly a long distance; this they can be considered as global pollination and uses some form of random walking. In addition, bees and birds may behave as Lèvy flight behavior [3], with jump or fly distance steps that obey a Lèvy distribution. Furthermore, flower constancy can be used as an increment step using the similarity or difference of two flowers [2].

6.1.2 The artificial flower pollination algorithm

Yang [1] idealized the characteristics of the pollination process, flower constancy, and pollinator behavior in the following rules [1]:

1. *Biotic and cross-pollination* are considered as global pollination processes with pollen carrying pollinators performing Lèvy flights.

2. *Abiotic and self-pollination* are considered as local pollination; mutation in differential evolution naming.

3. Flower constancy can be considered as the reproduction probability and is proportional to the similarity of two flowers involved.

4. Local pollination and global pollination are controlled by a switch probability $p \in [0, 1]$.

Due to the physical proximity and other factors such as wind, local pollination can have a significant fraction p in the overall pollination activities. There are two key steps in this algorithm: global pollination and local pollination.

In the *global pollination* step, flower pollens are carried by pollinators such as insects, and pollens can travel over a long distance. This ensures the pollination and reproduction of the most fit, and thus we represent the most fit as g_*. The first rule can be formulated as equation 6.1.

$$X_i^{t+1} = X_i^t + L(X_i^t - g_*) \tag{6.1}$$

where X_i^t is the solution vector i at iteration t and g_* is the current best solution. The parameter L is the strength of the pollination which is the step size randomly drawn from Lèvy distribution [3].

The *local pollination* (Rule 2) and flower constancy can be represented as in equation 6.2.

$$X_i^{t+1} = X_i^t + \epsilon(X_j^t - g_k^t) \tag{6.2}$$

where X_j^t and X_k^t are solution vectors drawn randomly from the solution set. The parameter ϵ is drawn from uniform distribution in the range from $[0, 1]$. We use a switch probability (Rule 4) or proximity probability p to switch between common global pollination to intensive local pollination. The FPA search can be summarized in the main procedure in algorithm (7).

6.2 Flower pollination algorithm variants

6.2.1 Binary flower pollination algorithm

In [4] a binary version of FPA algorithm was proposed to tackle for binary problems such as feature selection. In the proposed binary flower pollination algorithm the search space is modeled as a d-dimensional boolean lattice, in which the solutions are updated across the corners of a hypercube. Binary solution vector is used, where 1 corresponds to that feature which will be selected to compose the new dataset, with 0 being otherwise. Individual dimension continuous data is squashed using the formula 6.3 to to limit its range and hence can be threshold using the formula in 6.4.

$$S(X_i^j(t)) = \frac{1}{1 + e^{-X_i^j(t)}} \tag{6.3}$$

input : Search space
　　　　　Fitness function
　　　　　Number of flower agents
　　　　　Number of iterations (n)
　　　　　Switch probability p
output: Optimal Bat position and the corresponding fitness

　Initialize a population of n flowers/pollen gametes with random solutions.
Select the best solution called g_*
t=0;
;
while $t \leq n$ **do**
　　foreach *Flower i in the solution pool* **do**
　　if *rand* $<p$ **then**
　　　Draw a (d-dimensional) step vector L which obeys a Lèvy distribution.
　　　Apply Global pollination on solution i using equation 6.1.
　　else
　　　Draw ϵ from uniform distribution
　　　Randomly choose j, k solutions from the current pool of solutions
　　　Apply local pollination on solution i using j, k and equation 6.2
　　end
　　Evaluate the new solution if new solution is better, replace solution i with
　　it.
　　Endfor
　　Update the best solution.
end

Algorithm 7: Flower search algorithm (FSA)

where $X_i^j(t)$ is the *ith* dimension of the solution j at iteration t.

$$X_i^j(t) = \begin{cases} 1 \text{ if } S(X_i^j(t)) > \sigma \\ 0 \text{ otherwise} \end{cases} \tag{6.4}$$

where σ is a random number drawn from uniform distribution in the range $[0, 1]$.

6.2.2 Chaotic flower pollination algorithm

Chaotic maps are used in [5] to tune the flower pollination algorithm parameters and improve its performance. The main parameter that most affects the performance of FPA is the P parameter, which controls the switching between global and local pollination. Thus, in that work different chaotic maps are used to adapt this parameter and get enhanced performance. The chaotic maps used are

- Logistic map

- The sine map

- Iterative chaotic map

- Circle map

- Chebyshev map

- Sinusoidal map

- Gauss map

- Sinus map

- Dyadic map

- Singer map

- Tent map

6.2.3 Constrained flower pollination algorithm

A constrained version of the FPA is used in [6] where a penalty function is used to embed the constraints into the fitness function as follows:

$$f_{modified}(x) = f(x) + \lambda \sum_{n=1}^{K} Max(0, g_n) \tag{6.5}$$

where $f(x)$ is the original fitness function and *lambda* is the penalty coefficient, and K is the number of constraints and g_n is the constraint number n.

6.2.4 Multi-objective flower pollination algorithm

A multi-objective optimization problem with m objectives can be converted into a single fitness function simply by weighted summing of the individual objectives [7]. The fundamental idea of this weighted sum approach is that these weighting coefficients act as the preferences for these multi-objectives. For a given set of weights $(w_1; w_2; ...; w_m)$, the optimization process will produce a single point of the Pareto front of the problem. For a different set of w_i, another point on the Pareto front can be generated. With a sufficiently large number of combinations of weights, a good approximation to the true Pareto front can be obtained. It has proved that the solutions to the problem with the combined objective are Pareto-optimal if the weights are positive for all the objectives, and these are also Pareto-optimal to the original problem [7].

6.2.5 Modified flower pollination algorithm

In the work of [8] FPA is enhanced by enhancing the mutation process, dimension-by-dimension updating and local/global swtich probability adaptation. In the adapted mutation process, local neighborhood is used rather than the whole population. In this model, each vector uses the best vector of only a small neighborhood rather than the entire population to do the mutation. The neighborhood topology here is static and determined by the collection of vector subscript indices. And the local neighborhood model can be expressed in the following formula:

$$X_{i,G+1} = X_{i,G} + \alpha(X_{nBest_i,G} - X_{iG}) + \beta(X_{p,G} - X_{q,G}) \tag{6.6}$$

where $X_{nBest_i,G}$ is the best solution for vector $X_{i,G}$, p, q are solution indices in the neighborhood of $X_{i,G+1}$ and α, β are constants controlling the mutation.

Overall update evaluation strategy used in the standard FPA will affect the convergence rate and quality of solutions [8]. Thus, a dimension-by-dimension updating is proposed. Updating factors are applied in a dimension-by-dimension basis and only dimensions that cause enhancement in the fitness function are kept. In FPA, local search and global search are controlled by a switching probability $p \in [0, 1]$, and it is a constant value. Authors of [8] suppose that a reasonable algorithm should do more global search at the beginning of the searching process and global search should be less in the end. Thus, dynamic switching probability strategy (DSPS) to adjust the proportion of two kinds of searching processes was used. The switching probability is updated at each iteration of the optimization as follows:

$$p = 0.6 - 0.1 * \frac{T - t}{T} \tag{6.7}$$

where T is the number of iterations used and t is the current iteration number.

6.3 Flower pollination algorithm: hybridizations

In the literature, respectively few hybridizations are performed between FPA algorithm and other optimizers to enhance its performance in solving optimization tasks. This may be due to its good stand-alone performance and as it is respectively a new algorithm. Some of the motivations as mentioned in the literature are mentioned below.

- To improve the population diversity and to avoid the premature convergence of FPA

- To minimize the number of function evaluations that is a main problem in FPA

- To avoid falling in local minima

Table 6.1 states samples of hybridizations mentioned in the literature.

6.3.1 Flower pollination algorithm hybrid with PSO

In a proposed hybrid model in [6] the standard particle swarm optimization was applied until convergence, then the obtained solution became the initial solutions for the standard FPA that further enhances the obtained solutions. It is worth mentioning that the

Table 6.1: FPA hybridizations and sample references

Hybrid with	Target	Sample Reference
Particle swarm optimization (PSO)	To enhance premature convergence	[6]
Harmony search (HS)	To enhance premature convergence	[9]

used FPA is a chaotic one where the p parameter of the FPA is updated using a logistic map according to the following equation:

$$p_{n+1} = \mu p_n (1 - p_n) \tag{6.8}$$

where μ is set to 0.4, n is the iteration number, and p is the probability for switching between local and global pollination.

6.3.2 FPA hybridized with chaos enhanced harmony search

In [9] FPA population is initialized with the population at the convergence of the chaos enhance harmony search. Chaos logistic function is used to adapt the parameters of the harmony search optimizer.

6.4 Real world applications of the flower pollination algorithm

This section presents sample applications for flower pollination optimization. FPA was successfully applied in many disciplines in image processing domain, engineering, etc. The main focus is on identifying the fitness function and the optimization variables used in individual applications. Table 6.2 summarizes sample applications and their corresponding objective(s).

In [10] an automated retinal blood vessel segmentation approach based on flower pollination search algorithm (FPSA) was proposed. The FPA searches for the optimal clustering of the given retinal image into compact clusters under some constraints. The used clustering objective function was maximizing the clusters' compactness.

Economic load dispatch (ELD) is an optimization task in power system operation. ELD minimizing the fuel cost by optimally setting the real power outputs from generators is the objective of the ELD problem. In [11] FPA is used considering three different cost functions for minimizing the fuel cost. A multi-objective version of the FPA is proposed in [12].

FPA is used to solve the disc brake design problem. The objectives are to minimize the overall mass and the braking time by choosing optimal design variables: the inner radius r, outer radius R of the discs, the engaging force F, and the number of the friction

Table 6.2: CS applications and the corresponding fitness function

Application Name	*Fitness Function*	*Sample References*
Image segmentation	Divide image pixels into clusters and hence can be converted into bi-level image	[10]
Economic load dispatch (ELD)	Minimize the fuel cost by optimally setting the real power outputs from generators is the objective of ELD problem	[11]
Disc brake design problem	Minimize the overall mass and the braking time by choosing optimal design variables	[12]
Optimal reactive power dispatch problem	The objective of the reactive power dispatch is to minimize the active power loss in the transmission network	[13]
Antenna arrays directivity	Goal is to have high directivity, low side lobe level, and half-power beam width	[14]
Cluster head selection in wireless sensor network	Minimize power consumption and hence increase network lifetime	[15]
Transformer design problem	Achieve design objectives, namely, economically viable, has low weight, small size, and good performance	[16]
Integer programming	Find optimal solution of a given linear equation set with linear constraints	[5]

surface s, under design constraints such as the torque, pressure, temperature, and length of the brake.

A hybridization of flower pollination algorithm with chaotic harmony search algorithm (HFPCHS) is proposed to solve an optimal reactive power dispatch problem in [13]. The objective of the reactive power dispatch is to minimize the active power loss in the transmission network which can be mathematically described as follows:

$$f = \sum_{k \in Nbr} g_k(V_i^2 + V_j^2 - 2V_iV_j cos(\theta_{ij})) \qquad (6.9)$$

where g_k is the conductance of branch between nodes i and j, and Nbr is the total number of transmission lines in power systems.

In modern era antenna arrays are used for long distance communication. In wireless application, antenna array requires high directivity, low side lobe level, and half-power beam width. To aid the introduced geometry in order to produce a preconceived radiation pattern, evolutionary algorithm inspired from the FPA method was introduced [14].

The triangular prism, whose controlling factor is optimized by the flower pollination algorithm, steers for a beam pattern having side lobe level, half-power beam width, and directivity.

In the work in [15] FPA is used for multilevel routing in a wireless sensor network (WSN) to maximize network lifetime. The objective for the FPA is to select a set of cluster heads that are representative of the whole WSN and can provide data transmission to a base station. The selected cluster heads should be as close to all the WSN nodes as possible to minimize power consumed to transmit data from the sensors to the cluster head.

The aim of transformer design is to obtain the dimensions of all parts of the transformer in order to supply these data to the manufacturer. The transformer should be designed in a manner such that it is economically viable, has low weight, small size, and good performance, and at the same time it should satisfy all the constraints imposed by international standards. The problem of transformer design optimization is based on minimization or maximization of an objective function which is subjected to several constraints. Among various objective functions the commonly used objective functions are minimization of total mass, minimization of active part cost, minimization of main material cost, minimization of manufacturing cost, minimization of total owning cost or maximization of transformer rated power. The task of achieving the optimum balance between transformer performance and cost is complicated, and it would be unrealistic to expect that the optimum cost design would satisfy all the mechanical, thermal, and electrical constraints. FPA is proposed in [16] to tackle the transformer design problem.

Integer programming is NP-hard problems that are commonly faced in many applications [5]. It refers to the class of combinatorial constrained optimization problems with integer variables, where the objective function is a linear function and the constraints are linear inequalities. The linear integer programming (LIP) optimization problem can be stated in the following general form:

$$Max cx \tag{6.10}$$

where $x \in Z_n$ is a vector of n integers and c is a constant vector sized n, subject to

$$Ax \leq b \tag{6.11}$$

where A is an $m \times n$ matrix and m is the number of constraints. A chaotic version of FPA is used in [5] to tackle the discrete constrained optimization problem.

6.5 Flower pollination algorithm in feature selection

In this study FPA optimization is used to find an optimal feature subset in a wrapper-based manner. In wrapper-based feature selection a given classifier is used to guide the feature selection process; hence an intelligent searching method is always favored to minimize the time cost of search. The used classifier is the K-NN classifier [17] with K = 5 in all the experiments. Thus, the used fitness function of this task is formulated as

$$Wrapper = \alpha^* \gamma_R(D) + \beta^* \frac{|C - R|}{|C|} \tag{6.12}$$

where $\gamma_R(D)$ is the classification quality of condition attribute set R relative to decision D, $|R|$ is the number of a position or the length of selected feature subset, $|C|$ is the total number of features, α and β are two parameters corresponding to the importance of classification quality and subset length, $\alpha \in [0, 1]$, and $\beta = 1 - \alpha$.

Eighteen datasets from the UCI [18] repository are used for comparison; see Table 6.3. The eighteen datasets were selected to have various numbers of attributes and instances as representatives of various kinds of issues that the proposed technique will be tested on. For each dataset, the instances are randomly divided into three sets, namely, training, validation, and testing sets in cross-validation manner.

Each optimization algorithm is run for M times to test convergence capability for an optimizer. The used indicators used to compare the different algorithms are

- *Classification average accuracy*: This indicator describes how accurate the classifier is given the selected feature set. The classification average accuracy can be

Table 6.3: Description of the datasets used in experiments

Dataset	No. of Features	No. of Samples
Breastcancer	9	699
Exactly	13	1000
Exactly2	13	1000
lymphography	18	148
M-of-N	13	1000
Tic-tac-toe	9	958
Vote	16	300
Zoo	16	101
WineEW	13	178
spectEW	22	267
sonarEW	60	208
penglungEW	325	73
ionosphereEW	34	351
heartEW	13	270
congressEW	16	435
breastEW	30	569
krvskpEW	36	3196
waveformEW	40	5000

formulated as

$$AvgPerf = \frac{1}{M}\sum_{j=1}^{M}\frac{1}{N}\sum_{i=1}^{N} Match(C_i, L_i) \qquad (6.13)$$

where M is the number of times to run the optimization algorithm to select feature subset, N is the number of points in the test set, C_i is the classifier output label for data point i, L_i is the reference class label for data point i, and $Match$ is a function that outputs 1 when the two input labels are the same and outputs 0 when they are different.

- *Statistical mean* is the average of solutions acquired from running an optimization algorithm for different M running. Mean represents the average performance a given stochastic optimizer can be formulated as

$$Mean = \frac{1}{M}\sum_{i=1}^{M} g_*^i \qquad (6.14)$$

  where M is the number of times to run the optimization algorithm to select feature subset, and g_*^i is the optimal solution resulted from run number i.

- *Std* is a representation for the variation of the obtained best solutions found for running a stochastic optimizer for M different runs. Std is used as an indicator for optimizer stability and robustness, whereas Std is smaller. This means that the optimizer converges always to the same solution while larger values for Std mean many random results. Std is formulated as

$$Std = \sqrt{\frac{1}{M-1}\sum(g_*^i - Mean)^2} \qquad (6.15)$$

  where M is the number of times to run the optimization algorithm to select feature subset, g_*^i is the optimal solution resulted from run number i, and $Mean$ is the average defined in equation 6.14.

- *Average selection size* represents the saverage size of the selected features to the total number of features. This measure can be formulated as

$$AVGSelectionSZ = \frac{1}{M}\sum_{i=1}^{M}\frac{size(g_*^i)}{D} \qquad (6.16)$$

  where M is the number of times to run the optimization algorithm to select feature subset, g_*^i is the optimal solution resulted from run number i, $size(x)$ is the number of on values for the vector x, and D is the number of features in the original data set.

Table 6.4 outlines the average fitness value for the global best solution obtained by the different optimizers for the different 20 runs. We can see that the obtained best solution for almost all the datasets is minimum compare with the solutions obtained by both PSO or GA, which proves the capability of FPA to find an optimal solution

Table 6.4: Experiment results of mean fitness for the different methods

Dataset	FPA	GA	PSO
breastcancer	0.0258	0.0266	0.0266
breastEW	0.0295	0.0337	0.0337
congressEW	0.0317	0.0414	0.0386
exactly	0.0533	0.3060	0.2521
exactly2	0.2353	0.2293	0.2395
heartEW	0.1356	0.1533	0.1644
ionosphereEW	0.0974	0.1111	0.1077
krvskpEW	0.0339	0.0504	0.0645
lymphography	0.1429	0.1633	0.1673
m-of-n	0.0096	0.0868	0.0605
penglungEW	0.2333	0.2667	0.2500
sonarEW	0.1246	0.1594	0.1565
spectEW	0.1124	0.1281	0.1303
tic-tac-toe	0.2250	0.2419	0.2319
vote	0.0480	0.0540	0.0760
waveformEW	0.1990	0.2107	0.2147
WineEW	0.0000	0.0034	0.0136
zoo	0.0588	0.0831	0.1128

Table 6.5: Mean classification performance for the different optimizers over the different test data

Dataset	FPA	GA	PSO
breastcancer	0.9571	0.9571	0.9571
breastEW	0.9358	0.9326	0.9453
congressEW	0.9366	0.9200	0.9324
exactly	0.9483	0.6715	0.7213
exactly2	0.7375	0.7489	0.7387
heartEW	0.8022	0.8089	0.8133
ionosphereEW	0.8479	0.8308	0.8496
krvskpEW	0.9547	0.9335	0.9328
lymphography	0.7280	0.7320	0.7200
m-of-n	0.9910	0.8955	0.9297
penglungEW	0.6480	0.6480	0.6560
sonarEW	0.7400	0.6886	0.7171
spectEW	0.8270	0.7978	0.8112
tic-tac-toe	0.7461	0.7254	0.7342
vote	0.9260	0.9420	0.8960
waveformEW	0.7821	0.7656	0.7658
WineEW	0.9533	0.9300	0.9333
zoo	0.8845	0.9023	0.9030

Table 6.6: Standard deviation for the obtained best solution over the different runs

Dataset	FPA	GA	PSO
breastcancer	0.0068	0.0064	0.0056
breastEW	0.0127	0.0127	0.0142
congressEW	0.0079	0.0195	0.0159
exactly	0.0318	0.0075	0.0985
exactly2	0.0096	0.0155	0.0060
heartEW	0.0183	0.0165	0.0145
ionosphereEW	0.0318	0.0373	0.0281
krvskpEW	0.0110	0.0118	0.0359
lymphography	0.0250	0.0144	0.0171
m-of-n	0.0109	0.0549	0.0620
penglungEW	0.0697	0.0812	0.1062
sonarEW	0.0364	0.0271	0.0536
spectEW	0.0178	0.0293	0.0128
tic-tac-toe	0.0227	0.0303	0.0252
vote	0.0179	0.0167	0.0230
waveformEW	0.0050	0.0067	0.0066
WineEW	0.0000	0.0076	0.0142
Zoo	0.0690	0.0479	0.0684

Table 6.5 outlines tha average performance of the selected features by the different optimizers on the different datasets used. We can remark the advance of the features selected by FPA on the test data over the PSO and GA.

Table 6.5 outlines the obtained standard deviation for the fitness of the best solution obtained at the different 20 runs of every optimizer on every data sample. The obtained standard deviation value is minimum for FPA over most of the datasets used in the experiment, which proves the convergence capability of the algorithm that converges to the same/similar solution regardless of the initial random solutions at the begining of the optimization.

6.6 Chapter conclusion

This flower pollination chapter comes with deep technical discussion of the algorithm as one of the modern meta-heuristic optimizers. As a newly invented optimizer it still lacks of variants on the basic algorithm. Therefore, binary and chaotic, constrained versions of FPA were the first algorithms to be added to the basic FPA. This chapter also reviews the multi-objective flower pollination algorithm. A hybridization of flower pollination algorithm with particle swarm and harmony search was discussed. Successful applications of FPA were reviewed including from image segmentation, cluster head selection, and integer programming to economic load dispatch, disc brake design, and optimal reactive power dispatch problem. In addition, we discussed in more detail how flower pollination algorithm deals with the problem of feature selection and a multi-objective fitness function was designed to reflect the classification performance and number of selected features. It was tested and evaluated on a set of UCI standard datasets and proved good performance.

Bibliography

[1] Xin-She Yang, "Flower pollination algorithm for global optimization", in: *Unconventional Computation and Natural Computation* 2012, Lecture Notes in Computer Science, Vol. 7445, pp. 240-249, 2012.

[2] Xin-she Yang, Mehmet Karamanoglu, and Xingshi He, "Multi-objective Flower Algorithm for Optimization", *Procedia Computer Science*, 2013.

[3] I. Pavlyukevich, "Levy flights, non-local search and simulated annealing", *J. Computational Physics*, Vol. 226, pp. 1830-1844, 2007.

[4] Xin-She Yang, "Recent Advances in Swarm Intelligence and Evolutionary Computation", *Studies in Computational Intelligence*, Springer, Vol. 585, 2015.

[5] 1Ibrahim El-henawy and Mahmoud Ismail, "An improved chaotic flower pollination algorithm for solving large integer programming problems", *International Journal of Digital Content Technology and Its Applications* (JDCTA), Vol. 8(3), June 2014.

[6] O. Abdel-Raouf, M. Abdel-Baset, and I. El-henawy, "A new hybrid flower pollination algorithm for solving constrained global optimization problems", *International Journal of Applied Operational Research*, Vol. 4(2), pp. 1-13, 2014.

[7] X.S. Yang, M. Karamanoglu, and X.S. He, "Flower pollination algorithm: a novel approach for multiobjective optimization" *Engineering Optimization*, Vol. 46(9), pp. 1222-1237, 2014.

[8] RuiWang and Yongquan Zhou, "Flower Pollination Algorithm with Dimension by Dimension Improvement", *Mathematical Problems in Engineering*, Vol. 2014, Hindawi Publishing Corporation, Article ID 481791, 9 pages, http://dx.doi.org/10.1155/2014/481791.

[9] Osama Abdel-Raouf, Mohamed Abdel-Baset and Ibrahim El-henawy, "A novel hybrid flower pollination algorithm with chaotic harmony search for solving sudoku puzzles", *International Journal of Engineering Trends and Technology* (IJETT), Vol. 7(3), pp. 126-132, Jan 2014.

[10] E. Emary1, Hossam M. Zawbaa, Aboul Ella Hassanien, Mohamed F. Tolba, and Vaclav Snase, "Retinal vessel segmentation based on flower pollination search algorithm", in: 5th International Conference on Innovations in Bio-Inspired Computing and Application (IBICA2014), Czech Republic, 23-25 June 2014.

[11] R. Prathiba, M. Balasingh Moses, S. Sakthivel, "Flower pollination algorithm applied for different economic load dispatch problems", *International Journal of Engineering and Technology* (IJET), Vol. 6(2), pp. 1009-1016, April-May 2014.

[12] Xin-She Yang, Mehmet Karamanoglu, and Xingshi He, "Multi-objective flower algorithm for optimization", *Procedia Computer Science*, Vol. 18, pp. 861-868, 2013.

[13] K. Lenin, B. Ravindhranath Reddy, and M. Surya Kalavathi, "Shrinkage of active power loss by hybridization of flower pollination algorithm with chaotic harmony search algorithm", *Control Theory and Informatics*, Vol. 4(8), pp. 31-38, 2014.

[14] Surendra Kumar Bairwa, "Computer Aided Modeling of Antenna Arrays Interfaced with the Pollination Method", Department of Electrical Engineering, National Institute of Technology, Rourkela, India, http://ethesis.nitrkl.ac.in/5768/, 2014.

[15] Marwa Sharawi, E. Emary, Imane Aly Saroit, and Hesham El-Mahdy, "Flower pollination optimization algorithm for wireless sensor network lifetime global optimization", *International Journal of Soft Computing and Engineering* (IJSCE), Vol. 4(3), July 2014.

[16] H.D. Mehta and Rajesh M. Patel, "A review on transformer design optimization and performance analysis using artificial intelligence techniques", *International Journal of Science and Research* (IJSR), Vol. 3(9), pp. 726-733, September 2014.

[17] L.Y. Chuang, H.W. Chang, C.J. Tu, and C.H. Yang, "Improved binary PSO for feature selection using gene expression data", *Computational Biology and Chemistry*, Vol. 32, pp. 29-38, 2008.

[18] A. Frank and A. Asuncion, UCI Machine Learning Repository, 2010.

7 Artificial Bee Colony Optimization

7.1 Artificial bee colony (ABC)

A numerical optimization algorithm based on foraging behavior of honey bees, called artificial bee colony (or ABC), was proposed by Karaboga in [55].

7.1.1 Inspiration

Honey bees collect nectar from vast areas around their hive [2]. Bee colonies have been observed to send bees to collect nectar from flower patches relative to the amount of food available at each patch. Bees communicate with each other at the hive via a dance that informs other bees in the hive as to the direction, distance, and quality rating of food sources [2]. *Scout bees* are employed in locating patches of flowers, who then return to the hive and inform other bees about the goodness and location of a food source via a waggle dance. The scout returns to the flower patch with follower bees, *onlookers*. A small number of scouts continue to search for new patches, while bees returning from flower patches continue to communicate the quality of the patch.

Similar to other swarm-based optimization algorithms, it is important to establish a proper balance between exploration and exploitation in swarm optimization approaches. In the bee swarm optimization, different behaviors of the bees provide this possibility to establish a powerful balancing mechanism between exploration and exploitation. This property provides the opportunity to design more efficient algorithms in comparison with other population-based algorithms such as PSO and GA [3]. Spontaneous searching of the scout bees navigate them to explore new regions beyond that defined by employer or onlooker bees. Patterns by which the foragers and onlookers control their movements may provide good exploration approaches. For example, experienced foragers can use historical information about location and quality of food source to adjust their movement patterns in the search space.

7.1.2 Artificial bee colony algorithm

The ABC algorithm contains three types of bees: experienced forager, onlooker, and scout bees which fly in a D-dimensional search space to find the optimum solution. The bees are partitioned based on their fitness. The percentage of scout, forager, and experienced forager are usually determined manually.

The bees with the best fitness value are called *experienced foragers*; the bees with the worst fitness are called *scoutbees*, while the bees with mild fitness are called *onlooker bees*.

- *Scout bees:* In a bee colony, the food sources with poor qualities are abandoned and replaced by the new food sources that are found by the scout bees. The scout bees employ a random flying pattern to discover new food sources and replace the abandoned one with the new food source. A scout bee walks randomly in the search space and updates the position of the food sources if the new food source has better quality than previously found food source by that bee. The scout bee walks randomly in a region with radius t. The search region is centered at current position of the scout bee. So the next position of a scout bee is updated using equation 7.1.

$$\vec{x}_{new}(\vartheta, i) = \vec{x}_{old}(\vartheta, i) + RW(\tau, \vec{x}_{old}(\vartheta, i)) \tag{7.1}$$

where X_{old} is the abandoned solution, X_{new} is the new food source, and RW is a random walk function that depends on the current position of the *scout bee*, and the radius search ϑ.

- *Experienced bees:* The *experienced foragers* use their historical information about the food sources and their qualities. The information that is provided for an experienced forager bee is based on its own experience (or cognitive knowledge) and the knowledge of other experienced forager bees in the swarm. The cognitive knowledge is provided by experienced foragers. These bees memorize the decisions that they have made so far and the success of their decisions. An experienced forager i remembers the position of the best food source, denoted as $b(\ell, i)$, and its quality which is found by that bee at previous times. The position of the best food source is replaced by the position of the new food source if it has better fitness.

The social knowledge is provided by sharing the nectar information of the food sources which are found by experienced forager bees. All the experienced forager bees select the best food source that is found by the elite bee as their interesting area in the search space. The position of the best food source is represented as the vector $e(\ell, o)$. The *elite* bee is selected as the bee with best fitness. Since the relative importance of cognitive and social knowledge can vary from one cycle to another, random parameters are associated with each component of position update equation. So, the position of an experienced forager bee i is updated using equation 7.2.

$$\vec{x}_{new}(\xi, i) = \vec{x}_{old}(\xi, i) + w_b r_b(\vec{b}(\xi, i) - \vec{x}_{old}(\xi, i)) + w_e r_e(\vec{e}(\xi, o) - \vec{x}_{old}(\xi, i)) \tag{7.2}$$

where r_b and r_e are random variables of uniform distribution in range of $[0, 1]$ which model the stochastic nature of the flying pattern, the parameters w_b and w_e, respectively, control the importance of the best food source ever found by the *ith* bee and the best food source which is found by the elite bee, and $x_{new}(\ell, i)$

and $x_{old}(\ell, i)$, respectively, represent the position vectors of the new and old food sources which are found by the experienced forager i. The second component in the right side of the position update equation is the cognitive knowledge, which represents that an experienced forager is attracted toward the best position ever found by that bee. The third component is the social knowledge, which represents that an experienced forager is attracted towards the best position $e(\ell, o)$, which is found by the interesting elite bee.

- *Onlooker bees*: An *onlooker bee* uses the social knowledge provided by the experienced forager bees to adjust its moving trajectory the next time. At each cycle of the optimization, the nectar information about food sources and their positions (social knowledge) that are provided by the experienced forager bees are shared in the dance area. After that, an onlooker bee evaluates the provided nectar information, employs a probabilistic approach to choose one of the food sources, and follows the experienced forager bee which found the selected food source. In other words, an onlooker bee i selects an experienced forager j from a set of experienced bees as its own interesting elite bee denoted as $e(\ell, i)$ using the roulette wheel approach. The roulette wheel approach is used by onlooker bees for selecting their interesting elite bees. In this approach, as the quality of a food source is increased, the probability of its selection is increased, too. The flying trajectory of an onlooker bee i is controlled using equation 7.3.

$$\vec{x}_{new}(k, i) = \vec{x}_{old}(k, i) + w_e r_e (\vec{e}(\xi, i) - \vec{x}_{old}(k, i)) \qquad (7.3)$$

where $e(\ell, i)$ is the position vector of the interesting elite bee for onlooker bee i, $X_{old}(k, i)$ and $X_{new}(k, i)$, respectively, represent the position of the old food source and the new one that are selected by the onlooker bee i, and w_e, r_e probabilistically controls the attraction of the onlooker bee toward its interesting food source area.

The algorithm for ABC is outlined as shown in Algorithm 8.

7.2 ABC variants

7.2.1 Binary ABC

A chaotic fuzzy version of ABC was used in [18] for feature selection for classification purposes. The K-nearest neighbor (K-NN) classifier has been used in the fitness function to evaluate feature selection given a dataset. A binary version of the standard artificial bee colony was proposed in [4] for feature selection purposes. Initially, a set of food source positions (possible solutions) are randomly selected by the employed bees and their nectar amounts (fitness functions) are determined. One of the key issues in designing a successful algorithm for a feature selection problem is to find a suitable mapping between feature selection problem solutions and food sources in optimization algorithm. In the ABC optimization algorithm, the food sources are randomly generated. The solutions

input : *NScouts, NExperienced, NOnlooker* A user-defined Number of
scout, experienced and onlooker bees.
n Number of bees.
$Iter_max$ maximum allowed number of iterations.
output: Optimal bee position and its fitness

Generate initial population of n random positions
;
while *Stopping criteria not met* **do**
    Evaluate Individual bee position given the fitness function.
    Select the best position; Elitist, according to fitness values.
    Divide the swarm according to fitness of best into Experienced, Onlooker,
    Scout bees.
    foreach *ExperiencedBee$_i$* **do**
    Update the positions of experienced bees using equation 7.2 and given the
    current global best
    foreach *OnlookerBee* **do**
    Select one of the experienced bees as an elite one using Roulette wheel
    approach.
    Update the position of onlooker bees using equation 7.3.
    foreach *ScoutBee* **do**
    Walk randomly around the search space Update the position of scout bees
    according to equation 7.1.
end

Algorithm 8: Artificial bee colony optimization algorithm

should have values equal to 0 or to 1, where 0 denotes that the feature is not activated
and 1 denotes that the feature is activated. In the proposed algorithm, the food sources
are calculated exactly as in the initially proposed algorithm, and, then, the values are
transformed by using a sigmoid function.

$$sig(x_{ij}) = \frac{1}{1 + exp(x_{ij})} \quad (7.4)$$

where x_{ij} is the continuous position of bee i in dimension j, and then the food sources
are calculated by

$$y_{ij} = \begin{cases} 1 \text{ if } rand < sig(x_{ij}) \\ 0 \text{ otherwise} \end{cases} \quad (7.5)$$

where $x_{i}j$ is the continuous position of bee number i in dimension j, y_{ij} is the binary
transformed solution, and *rand* is a random number drawn from uniform distribution in
the range [0,1].

Afterward, the fitness of each food source is calculated and an employed bee is at-
tached to each food source. The employed bees return in the hive and perform the

waggle dance in order to inform the other bees (onlooker bees) about the food sources as in the continuous version of ABC.

Another variant of binary ABC was proposed in [5] that makes use of logical operators rather than mathematical ones. The reposition of onlooker and employed bees is updated as

$$V_i^j = X_i^j \bigoplus [\phi(X_i^j \bigoplus X_k^j)] \tag{7.6}$$

where V_i^j, X_i^j, X_k^j are the new, current, and random solutions in dimension j, $\bigoplus$ is the x/or logical operator, and ϕ is the not logical operator where the result of $X_i^j \bigoplus X_k^j$ is inverted if $\phi > 0.5$; otherwise it is not inverted.

7.2.2 Chaotic search ABC (CABC)

The local searching of ABC around the best solution is always limited and hence in [6] the chaotic search method is employed to solve this problem. In the CABC, onlooker bees apply chaotic sequence to enhance the local searching behavior and avoid being trapped into local optimum. For onlooker bees, chaotic sequence is mapped into the food source. Onlooker bees make a decision between the old food source and the new food source according to a greedy selection strategy. The chaos system used in this chapter is defined by equation 7.7.

$$x_{i+1} = \mu * x_i * (1 - x_i) \tag{7.7}$$

where μ is a chaotic attractor. If μ is equal to 4 then the above system enters into a fully chaos state, and x_{i+1} is the value of the variable x_i in i iteration. The new food source will be calculated as in equation 7.8.

$$x = x_{mi} + R * (2 * x_i - 1) \tag{7.8}$$

where x is the new food source; x_{mi} is the center of local search; selected bee position, x_i, is the chaos variable calculated as in equation 7.7; and R is the radius of the new food source being generated.

7.2.3 Parallel ABC

A message passing-base parallel version of the ABC was proposed in [7] to enhance its performance. The proposed work uses the static load program where they calculated the amount of assigned tasks required on each node, for the amount of assigned tasks on each node is proportional there to the computing capacity of the node. Thus, each node expects to complete the task at the same time, thereby minimizing the program run time.

The parallelized version was formulated as follows:

Step 1 Initialize the runtime environment.

Step 2 If the current process is the main process, get algorithms related to the initial parameters such as population size n, and send the above-mentioned parameters and split the sample to each slave process. Each process randomly generates the initial nectar location constituted of n solutions.

Step 3 Employ bees of each process in accordance with the formulas in the standard ABC search for new nectar, and calculate the position and moderation. If the new location is better than the original location, then replace the original location.

Step 4 Each process compares the number of iteration cycles with $maxcycle$, if $cycle > maxcycle$, then records the optimal solution; otherwise go to Step 3.

Step 5 Each process sends the results from the slave processes to the main process and then exits parallel environment and output the final best solution.

7.2.4 ABC for constrained problems

A version of the ABC for handling constrained optimization problems was proposed in [8]. The proposed method uses dynamic tolerance for equality constraints. To make easier for the ABC the satisfaction of equality constraints, a dynamic mechanism based on the tolerance value ϵ was employed. The tolerance value is defined as

$$
\epsilon = \begin{cases} 1 \text{ if } t = 1 \\ e^{(-(\frac{E(t).dec}{E_f}))} \text{ if } 1 < t < S \\ 0.0001 \text{if } t > S \end{cases} \tag{7.9}
$$

where $E(t)$ is the current number of evaluations performed, E_f is the total number of evaluations to be computed, and S is the evaluation number when the user wants ϵ become 0.0001 and dec is calculated as

$$
dec = \frac{9.21034 * E_f}{S} \tag{7.10}
$$

7.2.5 Lèvy flight ABC

Although ABC is better in setting the balance between exploration and exploitation compared to other algorithms, it has several deficiencies [9]. While it can perform better the global search by performing random searches around each food source and covering the search space optimally, it encounters several problems in the exploitation part, and gets stuck on local minimums particularly in complex multimodal functions [9]. If the original ABC algorithm fails to make improvement for a particular food source as much as a predetermined limit value number, then the employed bee having that source becomes a scout bee. Then this scout bee is randomly distributed within the search space, boundaries of which are determined. In this case, scout bees select a zone within the search space completely randomly, without using the results found until that moment as a result of the iterations. It is a very low probability for this newly selected food source to be better than the global best found until that moment. Scout bees find new food sources using Lèvy flight distribution with also the effect of global best instead of selecting a random food source. In this way, it is sought to attain a better solution using the best food source until that moment instead of the food source that could not be improved. By performing enough random searches, exploration of the ABC algorithm, which is already good in exploration, is improved. The scout bee repositioning is performed according to

$$X^{t+1} = X^t + \alpha \bigoplus L\grave{e}vy(\beta) \tag{7.11}$$

where α is the step size that should be related to the scales of the problem of interest and is selected randomly in this work.

7.2.6 Elitist ABC (E-ABC)

It was noticed that the ABC converges faster to a promising region of the search space, causing, sometimes, premature convergence. In order to improve its performance in constrained numerical search spaces, some modifications are added [8]. The expected effect of all changes is to slow down convergence by slowing down the replacement process of the food sources generated by the onlooker bees, modifying the operator of the employed bee and giving the scout bee more chances to generate good solutions in the neighborhood of the best solution so far. In the modified version the employed bees change their positions according to

$$V_{ij}^g = X_{ij}^g + 2\phi_i(X_{kj}^g - X_{ij}^g) \tag{7.12}$$

where V_{ij}^g is the new position, x_{ij}^g is the current position, X_{kj}^g is a randomly chosen solution from the current population, and ϕ_i is a constant in the range [0,1].

Another modification is used in the onlooker bee repositioning. Each onlooker bee generates a food source V_i^g near the best food source, so far X_B^g, and the replacement takes place only if the new food source V_i^g is better than that best solution in the population. The modified operator is formulated as

$$V_{ij}^g = X_{ij}^g + 2\phi_i(X_{ij}^g - X_{Bj}^g) \tag{7.13}$$

Scout bees are repositioned according to the following equation:

$$V_{ij}^g = X_{ij}^g + \phi_i(X_{kj}^g - X_{ij}^g) + (1 - \phi_i)(X_{Bj}^g - X_{ij}^g) \tag{7.14}$$

where X_{kj}^g is a randomly selected agent and X_{Bj}^g is the current global best solution. The aim of this modified operator is to increase the capabilities of the algorithm to sample solutions within the range of search defined by the current population.

7.2.7 Interactive artificial bee colony (IABC)

The original design of the onlooker bees' movement only considers the relation between the employed bee, experienced bees, which is selected by the roulette wheel selection, and the one selected randomly [10]. Newtonian law of universal gravitation described in equation 7.15 is used to the universal gravitations between the onlooker bee and the selected employed bees.

$$F_{12} = G * \frac{m_1 * m_2}{r_{21}^2} \hat{r_{21}} \tag{7.15}$$

F_{12} denotes the gravitational force heads from the object 1 to the object 2, G is the universal gravitational constant, m_1 and m_2 are the masses of the objects, r_{21} represents the separation between the objects, and $\hat{r_{21}}$ denotes the unit vector r_{21}. In the implementation the m1 is the fitness of the experienced bee selected by the roulette wheel, and m_2 represents the fitness function of the onlooker bee. The universal gravitation in equation 7.15 is formed in the vector format. Hence, the quantities of it on different dimensions can be considered independently. The updating of any of the onlooker bees is calculated as

$$x_{ij}^{t+1} = \theta_{ij}^t + F_{ikj}.[\theta_{ij}^t - \theta_{kj}^t] \tag{7.16}$$

where x_{ij}^t is the position of the updated onlooker bee at iteration t, θ_{ij}^t is the position of onlooker bee at time t, θ_{kj}^t is the position of the experienced bee, and F_{ikj} is calculated as in equation 7.15.

7.2.8 Pareto-based ABC

A Pareto-based enhanced version of ABC was proposed in [14]. An external Pareto archive set was introduced to record non-dominated solutions found so far. To balance the exploration and exploitation capability of the algorithm, the scout bees in the hybrid algorithm are divided into two parts. The scout bees in one part perform random search in the predefined region while each scout bee in another part randomly select one non-dominated solution from the Pareto archive set.

7.2.9 Fuzzy chaotic ABC

Chaotic ABC based on the fuzzy system was proposed in [18]. The fuzzy logic is used for ambiguity removal while chaos is used for generating better diversity in the initial population of bee colony optimization algorithm. The chaos theory has been incorporated for three main purposes:

- Increasing diversity in the initial population

- Finding the neighborhood around a food source

- Generating random numbers

The logistic chaos map was selected as the one for chaotic number generation.

The decision for assigning a bee to one of the three classes, namely, scout, onlooker, and experience, is based on fuzzy membership function with a single parameter: fitness. A Gaussian rule base with Gaussian membership function was manually set and used for bee classification. Further the standard ABC optimization based on the chaotic map and fuzzy bee classification is used for the optimization.

7.2.10 Multi-objective ABC

A multi-objective version of the ABC optimizer (MO-ABC) was proposed in [21]. Multi-objective optimization entails finding the best optimal solution from all numbers of objectives simultaneously. The multi-objective algorithm (MOABC) method was proposed to find a set of non-dominated solutions that are known as the Pareto optimality. A solution X_1 dominates another solution X_2 if the following was satisfied:

$$\forall_{i \in \{1,2,...,m\}} \ f_i(X_1) \leq f_i(X_2) \tag{7.17}$$

$$\exists_{j \in \{1,2,...,m\}} \ f_i(X_1) < f_i(X_2) \tag{7.18}$$

In the MO-ABC the non-dominated solutions are stored in an external archive. The employed bee comes into the hive with a non-dominate solution to go into the archive. After all employed bees finish their search processes, they insert all new non-dominated solutions into archive. Onlooker bees will select an archive member proportional to the quality of food source with the roulette wheel selection, and pick the leader and update it. When all the different types of bees (i.e., employed, onlooker, and scouts) are evaluated, the archive is filled with the non-dominated bees. The content of the external archive has a bounded size. If the archive is full, the archive gets updated at the end of each iteration.

7.2.11 JA-ABC

Despite its excellent performance, ABC suffers from slow convergence speed and premature convergence tendency. To enhance the convergence speed of the standard ABC [12], a ratio of all possible solutions, which have the lowest fitness value, are to be updated. The proposed modification is to update poor possible solutions using mutation around the best solution; see equation 7.19.

$$Z_{ij}^t = best^t + \phi(y_p^t - y_k^t) \tag{7.19}$$

where Z_{ij}^t is the new solution, $best_t$ is the best solution found ever, y_p and y_k are two random agents drawn from the bee agents, and ϕ is a random number drawn from uniform distribution in the range [-1,1].

7.3 ABC hybridizations

In the literature many hybridizations are performed between ABC algorithm and other optimizers to enhance its performance in solving optimization tasks. Some of the motivation as mentioned in the literature are mentioned below.

- To improve the population diversity and to avoid the premature convergence of FFA

- To minimize the number of function evaluations that are a main problem in FFA

- To avoid falling in local minima

Table 7.1 states samples of hybridizations mentioned in the literature.

Table 7.1: ABC hybridizations and sample references

Hybrid with	Target	Sample Reference
Least squares method (LS)	Enhances the convergence speed by employing LS to optimize linear portions of the search space	[19]
Differential evolution (DE)	Avoids premature convergence of DE and stagnation and to avoid convergence to a local minima	[11]
Quantum evolution (QE)	Improves the convergence speed and preserve the instructive information; also improves the convergence speed as well as increases the diversity of the population	[13]
Particle swarm optimization (PSO)	Enhances the convergence rate of the hybrid algorithm while the explorative power of ABC enhances space searching	[15]
Levenberq-Marquardt	Exploits the fast convergence of Levenberq-Marquardt to initialize ABC to near optimal solution	[17]
Harmony search (HS)	Tolerates the premature and/or false convergence, slow convergence of harmony search especially over multi-modal fitness landscape	[20]
Ant colony optimization (ACO)	Exploits the simplicity and robustness of ACO. The search capability of the ABC algorithm is used for escaping from the local minimum solution. The hybrid algorithm is able to quickly find the correct global optimum. The performance of algorithm is improved by dividing the optimization problem into continuous and discrete category stages; thus, it decreases the size of the search space. The hybrid algorithm is able to overcome the drawback of classical ant colony algorithm which is not suitable for continuous optimizations	[21]

7.3.1 Hybrid ABC and least squares method

The proposed approach based on hybrid ABC optimization and weighted least squares (LS) method was proposed in [19]. In some search spaces such as Takagi-Sugeno (TS)-type model, part of the unknowns is linearly related to the output fitness and another variable set is nonlinearly related to the fitness. In the work [19] ABC optimization was used to optimize a set of nonlinearly related unknowns and given the ABC guess an LS method is used to find the exact values of the rest of the variables and so on. Thus the ABC is used for the nonlinear portions while LS is used to calculate the nonlinear portions of the optimization problem and hence ensures high performance in both accuracy and performance.

7.3.2 Hybrid differential artificial bee colony algorithm

In [11] a modified version of ABC is presented. In some problems, differential evolution (DE) may stop proceeding toward the global optimum: stagnation. DE also suffers from premature convergence, where the population converges to some local optima of a multimodal objective function, losing its diversity. Like other evolutionary computing algorithms, the performance of DE deteriorates with the growth of the dimensionality of the search space as well. The advantage of nature-inspired heuristic global search algorithms is less likely to be entrapped in local optima, but the convergence rate will slow down and the computational complexity is high at a later stage. The main goal in [11] is to make hybrid version of the ABC with DE to combine their advantages.

Pipelining-type hybrid method is employed in this case because of its advantages. In the pipelining hybridization, the optimization process is applied to each individual in the population, followed by further improvement using DE search. According to this method every generation, after the ABC is applied to all individuals in the population, select n best differential vectors from the current population based on the fitness values to generate the initial population required for local search via DE. This search continues until it reaches maximum number of generations or it satisfies predefined criteria. In brief the mutation, crossover, and selection operators are applied on each agent from the ABC population in each iteration.

7.3.3 Quantum evolutionary ABC

The information in a quantum chromosome is more than that in a classical chromosome [13]; the population size is decreased and the diversity is improved. In the quantum evolution (QE) the mutation operation is not totally random but directed by some rules to make the next generation better and increase the speed of convergence. The crossover operation can avoid premature convergence and stagnation problems. The quantum evolution algorithm (QEA) still has scope for improvement. QEA could not always reach the best solution of the problem meaning that it has a considerable probability of premature convergence. To make use of the advantages and tolerate for the disadvantages the hybrid QDE and ABC are proposed in [13]. The main change in the ABC is in the employer bee updating. The updating equation is as mentioned in the experience bee updating and followed by steps of quantum crossover and quantum mutation. Single point crossover hired from QDE is used as well as the quantum rotation. In the single point crossover process, a roulette selection operation is used to choose two quantum chromosomes from the parent generations, and then the child generation is produced by crossover. The result of crossover and the original solution are compared and the one with best fitness is kept. This operation is mainly to improve the convergence speed and preserve the instructive information. The selected solution is used as the mutation director and implements the quantum mutation operation. This operation is also to improve the convergence speed as well as to increase the diversity of the population. After all the operations, finally the employed bee selects the better population as the new source position to remember.

7.3.4 ABC hybrid with PSO

A hybrid version of ABC with particle swarm optimization (PSO) was proposed in [15]. The algorithm starts by building random initial personal best positions for particles/bees. Then in the main loop it improves the personal bests using successive application of PSO followed by ABC updating. When the main loop ends, the algorithm scans the latest personal best positions and returns the best of them as the global best. The velocity concept employed by PSO enhances the convergence rate of the proposed hybrid algorithm while the explorative power of ABC enhances space searching.

7.3.5 ABC hybrid with Levenberq–Marquardt

A hybridization between ABC and Levenberq-Marquardt has been proposed in [17] for optimal searching for neural network parameters. The Levenberq-Marquardt proves high speed was exploited to initialize the ABC population which further used ABC principles to reach an optimal solution. The exploration as well as exploitation capabilities of ABC were well employed in conjunction with the fast convergence capabilities of Levenberq-Marquardt to reach optimal fast solutions.

7.3.6 ABC hybrid with harmony search

Experiments with the harmony search (HS) meta-heuristics over the standard numerical benchmarks suggest that the algorithm does suffer from the problem of premature and/or false convergence, slow convergence especially over multi-modal fitness landscape [20]. The harmony memory plays a key role in the HS algorithm. To improve the convergence of the HS, ABC is used to optimize the harmony memory.

The ABC is applied to optimize harmony memory as a learning mechanism. The harmony memory has considered food sources, and is explored and exploited by the employed bees, the onlooker bees, and the scout bees. Then, a new harmony vector is generated [20].

7.3.7 ABC hybrid with ant colony optimization(ACO)

Ant colony optimization (ACO) has a population-based search capability as well as simplicity and robustness in different optimization tasks. ACO used a heuristic technique to produce a good initial solution and determine a good search direction depending on the experience. It is worthy of note that this strategy often helps ACO find a good solution and it causes ACO to be trapped in local minima.

To overcome these disadvantages and integrate the merits of both ACO and ABC, the hybrid ACOABC algorithm is introduced in [21]. Hybridizing ACO with ABC algorithm takes advantage of ACO and ABC by dividing the optimization problem into two parts. In each iteration, ant builds candidate solutions that are further optimized by ABC. The proposed algorithm has several characteristic features. First, the search capability of the ABC algorithm is used for escaping from the local minimum solution. This causes ABC to generally affect the search process since it is able to quickly find the correct global optimum. Second, the performance of algorithm is improved by dividing the optimization problem into continuous and discrete category stages; thus, it decreases the size of the search space. Finally, it is able to overcome the drawback of classical ant colony algorithm which is not suitable for continuous optimizations [21].

To avoid falling in a local minima, ACO is enhanced using the local searching power of FFA [22]. The pheromone generated by the ACO is converted into candidate solutions using equation 7.20.

$$p_{ij}^k(t) = \begin{cases} \frac{[\tau_{ij}(t)]^\alpha}{\sum_{l \in allowed_k}[\tau_{il}(t)]^\alpha} & \text{for all } x_{ij} \text{ , } x_{ij} \in allowed_k \\ 0, & otherwise \end{cases} \qquad (7.20)$$

where $x_{ij}^k(t)$ is probability of that option, x_{ij} is chosen by ant k for variable i at time t, and $allowed_k$ is the set of the candidate values contained in the group i.

The candidate ACO solutions are used to initialize the FFA for further enhancement. The random component in the standard FFA updating equation is adapted as follows:

$$\alpha_{t+1} = \alpha_t \theta^{1-\frac{t}{Tc}} \qquad (7.21)$$

where t is iteration number, Tc maximum number of iterations, and $\theta \in [0,1]$ is the randomness reduction constant.

7.4 Artificial bee colony in real world applications

This section reviews some success applications that used ABC. FFA was successfully applied in many disciplines in decision support and decision making, in the engineering field, computer science, communication, image processing, and enhancement. The wide range of ABC applications may be interpreted by its capability to quickly converge to optimal solution thanks to the balance between exploration and exploitation. Our main focus in this section is to highlight on identifying the fitness function and the optimization variables used in individual applications. Table 7.2 summarizes sample applications of ABC and their corresponding objective(s).

ABCs have been employed to solve several problems in image processing areas. ABCs are used in [23] to discriminate between noisy and normal regions. A noise probability threshold is required to be estimated using ABC. ABC optimization is used before performing the FCM clustering algorithm.

The work in [24] presents an automatic *image enhancement* method based on ABC algorithm. In this method, ABC algorithm is applied to find the optimum parameters of a transformation function, which is used in the enhancement by utilizing the local and global information of the image. In order to solve the optimization problem by ABC algorithm, an objective criterion in terms of the entropy and edge information is introduced to measure the image quality to make the enhancement as an automatic process. Several images are utilized in experiments to make a comparison with other enhancement methods, which are genetic algorithm-based and particle swarm optimization algorithm-based image enhancement methods.

In [25], a new approach to automatic image enhancement using ABC is implemented by specifying intensity of the edge pixels and also earlier reported PSO results were used. Further comparativity analysis is performed between ABC and PSO results. The obtained results indicate that the proposed ABC yields better results in terms of both the maximization of number of pixels in the edges and pick signal-to-noise ratio (PSNR).

ABC algorithm is used in medical *image retrieval* [26]. Where ABC is used to optimize multi-feature similarity, score fusion allows for reasonable retrieval rate with less retrieval time and high classification accuracy.

The work in [27] proposed a hybrid swarm optimization technique to outperform the individual performance of ABC and PSO. The experimentation was performed in the content-based image retrieval (CBIR) to evaluate the performance of the proposed hybrid algorithm, and the hybrid one proved superior performance.

ABC is used in [28] in medical *image registration*. ABC can obtain a set of optimized parameters for registration with the largest similarity measure. By applying the proposed ABC method to medical images, the experimental results showed that the method achieved better accuracy in registration than the conventional optimization methods that are commonly used in medical image registration. ABC is also applied for image registration in [29] and is compared to PSO and proves good performance.

ABC algorithm is used to improve the efficiency of FCM on abnormal brain segmentation [30]. ABC is used to minimize the fuzzy c-means objective function to achieve better segmentation results.

Table 7.2: ABC applications and the corresponding fitness function

Application Name	Fitness function	Sample References
Image Enhancement	Discriminate noisy and noise free regions	[23]
Image Enhancement	Intensity transformation function parameter identification	[24]
Image Enhancement	Find Edge pixels with maximum signal to noise ratio	[25]
Medical Image Retrieval	Optimize multi-feature similarity score fusion to allow for reasonable retrieval rate with less retrieval time and high classification accuracy	[26, 27]
Image Registration	Find a set of optimized registration parameters with the largest similarity measure	[28, 29]
Image Segmentation-brain images	Minimize the fuzzy c-means objective function	[30, 31]
Medical image segmentation	Find Gaussian mixture model parameter maximizing data fit	[32]
Image Segmentation	Optimize Otsu's objective function	[33]
Image Segmentation	Maximize Support vector machine performance	[34]
Brain tissue segmentation	Multi-threshold selection to maximize intra-cluster homogeneity	[23, 35]
Image Segmentation	Maximize Tsallis entropy	[36]
Remote sensing image segmentation	Maximize gray level entropy	[37]
Multi-level thresholding	To produce the best compressed image in terms of both compression ratio and quality	[39–42, 51]
Multi-temporal image fusion	Maximize entropy	[43]
Multi-temporal image fusion	Maximize the similarity score based on color and texture features	[44]
Shape Matching	Optimize edge potential function	[32, 45, 46]
Solve the travelling sales man problems	Select route with minimum cost	[47]
Capacitated vehicle routing problem	Find best route for a set of vehicles with limited capacity	[48]
Job scheduling	To minimize overall work load, completion time and machine work load	[14]
Garlic Expert Advisory System	Identify the diseases and disease management in garlic crop production to advise the farmers	[54]
subway routes selection	Aims to maximize the population covered by subway routes	[49]
Multiple sequence alignment	Maximize sequence correlation	[50]
Neural network training	Find neural weights and biases that output minimum square error	[51–53, 55–57, 59–61]
Fuzzy system training	Estimation of membership function parameters minimizing total mean squares error	[16, 19]
Feature selection	Rough set classification performance	[62]

Table 7.2: ABC applications and the corresponding fitness function (continued)

Application Name	*Fitness function*	*Sample References*
Initialize Fuzzy system training	Clustering using fuzzy C-means	[51, 63]
Train least squares support vector machine	Find optimal neural weights maximizing neural performance	[60]
Power system stabilizer	Improve the system performance and enhance power system stability	[64]
Feature Selection	Maximizing the performance of rough set	[62, 65, 66]
Data clustering	Optimal partition of data point with guidance of Debs rules	[67]
PID controller design	Goal is to improve the Automatic Voltage Regulator system step response characteristics	[6]
Wire electrical discharge machining	Modelling and optimization of process parameters of wire electrical discharge machining	[68]
capacitor placement in distribution systems	The objective is improving the voltage profile and reducing power loss	[72]
automatic voltage regulator (AVR) system	Improving the performance of the controller	[73]
Deployment of wireless sensor network	Achieve better performance and much coverage	[75–77]

ABC is used in [32] for selecting gaussian mixture model parameters of the histogram of medical image. ABC optimization shows fast convergence and low sensitivity to initial point selection than expectation the maximization method which is common. A combination of the 2D Otsu method with a modified ABC algorithm (called adaptive ABC or AABC) to reduce the response and computational time for image segmentation of pulmonary parenchyma is proposed [33].

Multisliced CT images are segmented using ABC algorithm. CT segmentation can be modeled as a nonlinear multi-modal global optimization problem. In [34] SVM along with ABC are used for segmenting tumor from a CT/MRI image. It reduces the response time and computational time. It is efficient in terms of performance, speed, and avoidance trapping in local minima points. In this chapter the ABC algorithm classified all pixels into two groups based on the noise probability; the two groups are Normal and Noisy.

Fuzzy inference strategy was embedded into the ABC system to construct a segmentation system named fuzzy artificial bee colony system (FABCS) in [31]. They set a local circular area with a variable radius by using a cooling schedule for each bee to search suitable cluster centers with the FCM algorithm in an image. The cluster centers can be calculated by each bee with the membership states in the FABCS and then updated iteratively for all bees in order to find near-global solution in MR image segmentation.

Study in [23] shows the implementation of artificial bee algorithm for MRI fuzzy segmentation of brain tissue, where ABC is used for multi-threshold segmentation of images. Synthetic MRI data are segmented using ABC in [35] and compared with other well-known segmentation methods such as k-means clustering and PSO optimization. The experimental results have demonstrated that ABC-based clustering algorithm gives more accurate outcomes than the other clustering methods.

The work in [36] uses the ABC approach since execution of an exhaustive algorithm would be too time-consuming for threshold selection. The experiments demonstrate that (1) the Tsallis entropy is superior to traditional maximum entropy thresholding, maximum between class variance thresholding, and minimum cross entropy thresholding; (2) the ABC is more rapid than either genetic algorithm or particle swarm optimization.

A system for segmentation of synthetic aperture radar (SAR) images is presented in [37]. Due to the presence of speckle noise, segmentation of SAR images is still a challenging problem. A fast SAR image segmentation method based on ABC algorithm is proposed in [37]. In this method, threshold estimation is regarded as a search procedure that searches for an appropriate value in a continuous gray-scale interval. Hence, ABC algorithm is introduced to search for the optimal threshold. In order to get an efficient fitness function for ABC algorithm, after the definition of gray number in Grey theory, the original image is decomposed by discrete wavelet transform. Then a filtered image is produced by performing a noise reduction to the approximation image reconstructed with low-frequency coefficients. At the same time, a gradient image is reconstructed with some high-frequency coefficients. A co-occurrence matrix based on the filtered image and the gradient image is therefore constructed, and an improved two-dimensional gray entropy is defined to serve as the fitness function of ABC algorithm.

A global multilevel thresholding method for image segmentation which is based on ABC approach was proposed in [51]. ABC was used to determine the thresholds to produce the best compressed image in terms of both compression ratio and quality [39,40]. A new multilevel maximum entropy thresholding algorithm based on the technology of ABC and the experimental results demonstrated that the proposed algorithm can search for multiple thresholds that are very close to the optimum ones examined by the exhaustive search method [41]. In [43] ABC is used for optimal fusion of multitemporal images. The work proposed makes use of entropy as a fitness function for the fused image and the input images are images acquired at two different dates.

ABC was used in [44] for multitemporal image fusion. ABC algorithm is used to fuse a similarity score based on color and texture features of an image thereby achieving very high classification accuracy and minimum retrieval time.

ABC has been applied to the object recognition in the images to find a pattern or reference image (template) of an object [45]. ABC employed for the problem of handwritten digits recognition where it has been used to train a classifier [46]. A novel shape-matching approach to visual target recognition for aircraft at low altitude where ABC is employed to optimize edge potential function was used for automatic detection of multiple circular shapes on images that considers the overall process as a multimodal optimization problem [32].

The paper [47] makes use of ABC to solve the standard traveling salesman (TSP) problem. In TSP a number of cities have to be visited by a salesman who must return to the same city with the solution of shorter routes. A salesman travels around a given set of cities, and return to the beginning of the path (from where he started), covering the smallest total distance. The traveling sequence has to comply with a constraint, that is, the salesman will start at a city, visit each city exactly once, and back to the start city. The resulting route should incur a minimum cost. Thus, the fitness function is the cost of the salesman's path.

The paper [48] introduces an ABC heuristic for solving the capacitated vehicle routing problem which is an NP-hard problem. The capacitated vehicle routing problem (CVRP) is defined on a complete undirected graph G = (V,E), where V = 0,1,..., n is the vertex set and E = (i, j): i, j $\in$ V, i ¡ j is the edge set. Vertices $1, ..., n$ represent customers; each customer i is associated with a nonnegative demand d_i and a nonnegative service time s_i. Vertex 0 represents the depot at which a fleet of m homogeneous vehicles of capacity Q is based. The fleet size is treated as a decision variable. Each edge (i, j) is associated with a nonnegative traveling cost or travel time Cij. The CVRP is to determine m vehicle routes such that (a) every route starts and ends at the depot, (b) every customer is visited exactly once, (c) the total demand of any vehicle route does not exceed Q, (d) the total cost of all vehicle routes is minimized, In some cases, the CVRP also imposes duration constraints where the duration of any vehicle route must not exceed a given bound.

Flexible job scheduling problem (FJSP) considers n jobs to be processed on m machines as in the standard job scheduling problem [14]. There are some assumptions and constraints that are common in the FJSP such as

1. Each job has a predefined number of operations and a known determined sequence among these operations.

2. Each machine and each operation are ready at zero time.

3. Each machine can only process one operation at a time, and each job must be processed on one machine at a given time.

4. Each machine can process a new operation only after completing the predecessor operation.

5. Each operation can be operated on a given candidate machine set instead of only one machine like in JSP.

6. Given an operation O_{ij} and the selected machine M_k, the processing time p_{ijk} is also fixed.

The common objectives for this problem are

- To minimize maximum completion time

- To minimize total work load

- To minimize critical machine work load

In [14] a pareto-based ABC was used to find the optimum for the FJSP with the above-mentioned constraints, assumptions, and objectives.

A new "Garlic Expert Advisory System" based on ABC, aims to identify the diseases and disease management in garlic crop production to advise the farmers in the villages to obtain standardized yields [54].

A method based on ABC is proposed to locate the subway routes that aim to maximize the population covered by subway routes [49]. Multiple sequence alignment (MSA) is an important method for biological sequence analysis and involves more than two biological sequences, generally of the protein, DNA, or RNA type. This method is computationally difficult and is classified as an NP-hard problem.

A parallel version of ABC is used in the optimization and investigation of multiple alignments of biological sequences, which is highly scalable and locality-aware [50]. The method is iterative and the concept is of algorithmic and architectural space correlation.

ABC was used in [51] to train a *classifier* of forward neural network to classify MR brain images. The paper [52] presents an intelligent computer-assisted mass classification method for breast DCE-MR images. It uses the ABC algorithm to optimize the neural network performing benign-malignant classification on the region of interest. A three-layer neural network with seven features was used for classifying the region of interest as benign or malignant. The network was trained and tested using the ABC algorithm and was found to yield a good diagnostic accuracy.

In [53], case-based reasoning (CBR) is used to describe a physician's expertise, intuition, and experience when treating patients with well-differentiated thyroid cancer.

Various clinical parameters (the patient's diagnosis, the patient's age, the tumor size, the existence of metastases in the lymph nodes, and the existence of distant metastases) influence a physician's decision-making in dose planning. The weights (importance) of these parameters are determined here with the bee colony optimization (BCO) meta-heuristic.

In [55] authors employed ABC for training feed-forward neural networks, i.e., searching optimal weight set. ABC applied ABC on training feed-forward neural networks to classify different datasets that are widely used in the machine learning community [56]. ABC employed to train a multi-layer perception neural network that is used for the classification of the acoustic emission signal to their respective source [57]. A comparison of RBF neural network training algorithms was made for inertial sensor-based terrain classification [59]. ABC used to train neural network for bottom hole pressure prediction in under-balanced drilling [58], an integrated system where wavelet transforms and recurrent neural networks based on ABC (called ABC-RNN) are combined for stock price forecasting [60]. An ABC-based methodology maximizes its accuracy and minimizes the number of connections of an artificial neural network by evolving at the same time the synaptic weights, the artificial neural networks architecture, and the transfer functions of each neuron [61].

A hybrid Levenberq-Marquardt and ABC optimizer has been used in [17] for training of a feed-forward neural network for classification purposes.

Fuzzy logic is conceived as a better method for sorting and handling data; also it has proven to be an excellent choice for many control system applications due to mimicking human logic. It uses an imprecise but very descriptive language to deal with input data, more like a human operator [16]. The key component of a fuzzy system is the membership function for defining the fuzzy sets and the rule base for producing the output. A fuzzy membership parameter selection is a key factor for the performance of the fuzzy system. ABC has been used in [16] for selection of membership parameters for construction of a fuzzy system.

A hybrid version of least-squares estimation and ABC has been used in [19] for estimation of fuzzy model parameters. The parameters to be optimized are

- The first part deals with the rule selection and the optimization of the number of rules.

- The second part deals with the optimization of rule premise fuzzy sets.

- The third part deals with the optimization of the rule-consequent parameters.

A new methodology was used for automatically extracting a convenient version of T-S fuzzy models from data using a novel clustering technique, called variable string length ABC algorithm based fuzzy c-means clustering approach [51]. Studying the risk of dams in the perspective of clustering analysis and to improve the performance of fuzzy c-means clustering, they proposed an ABC with fuzzy c-means [63].

A concept for machine learning integrates a grid scheme into a least-squares support vector machine (called GS-LSSVM) for classification problems, where ABC is used to optimize parameters for LSSVM learning [60].

The work in [62] proposes a new feature selection method based on rough set theory hybrid with BCO in an attempt to combat this. This proposed work is applied in the medical domain to find the minimal reducts and experimentally compared with the quick reduct, entropy-based reduct, and other hybrid rough set methods such as genetic algorithm (GA), ant colony optimization, and particle swarm optimization (PSO).

An improved rough set based attribute reduction (RSAR), namely, independent RSAR hybrid with ABC algorithm, finds the subset of attributes independently based on decision attributes (classes) at first and then finds the final reduct [65]. The use of the ABC algorithm as a new tool for data mining, particularly in classification tasks, indicated that ABC algorithm is competitive, not only with other evolutionary techniques but also to industry standard algorithms such as self-organizing maps (SOM), naive bayes, classification tree, and nearest neighbor (kNN) [66].

An ABC clustering algorithm optimally partitions n objects into k clusters where Debs rules are used to direct the search direction of each candidate [67].

A modified version of the ABC has been used in [6] for properly tuning the parameters of PID controller, because many industrial plants are often burdened with problems such as high orders, time delays; and nonlinearities and the fact that the mathematical model of plant is difficult to determine.

ABC has found several applications in mechanical and civil engineering areas. ABC employed for modeling and optimization of process parameters of wire electrical discharge machining [68]. A modification on the original ABC algorithm was to optimize the equilibrium of confined plasma in a nuclear fusion device and its adaption to a grid computing environment [69]. An ABC algorithm was introduced for structural optimization of planar and space trusses under stress, displacement, and buckling constraints [70].

ABC was used by some researchers to solve the optimization problems encountered in electrical engineering. A new method based on ABC algorithm applies for determining the sectionalizing switch to be operated in order to solve the distribution system loss minimization problem [68]. An ABC algorithm was used for simultaneous coordinated tuning of two power system stabilizers to damp the power system inter-area oscillations [71].

ABC applied for optimal control of flexible smart structures bonded with piezoelectric actuators and sensor and the optimal locations of actuators/sensors and feedback gain are obtained by maximizing the energy dissipated by the feedback control system [68].

The optimal tuning for the parameters of the power system stabilizer, can improve the system damping performance within a wide region of operation conditions to enhance power system stability using ABC [64].

A new method applies ABC for capacitor placement in distribution systems with an objective of improving the voltage profile and reduction of power loss [72].

A study on comparative performance analysis of ABC for automatic voltage regulator (AVR) systems showed that the ABC algorithm can be successfully applied to the AVR system for improving the performance of the controller [73].

A few applications of ABC related to wireless sensor networks were found in the literature. The use of ABC for the sensor deployment problem which is modeled as a data clustering problem was proposed in [75]. ABC-based dynamic deployment approach for stationary and mobile sensor networks was to achieve better performance by trying to increase the coverage area of the network [76]. ABC applied to the dynamic deployment of mobile sensor networks to gain better performance by trying to increase the coverage area of the network [77].

7.4.1 Artificial bee colony in retinal vessel segmentation

Accurate segmentation of retinal blood vessels is an important task in computer- aided diagnosis and surgery planning of retinopathy. Despite the high resolution of photographs in fundus photography, the contrast between the blood vessels and retinal background tends to be poor. Furthermore, pathological changes of the retinal vessel tree can be observed in a variety of diseases such as diabetes and glaucoma. Vessels with small diameters are much liable to effects of diseases and imaging problems. In this case study, an automated retinal blood vessel segmentation approach based on ABC optimization is discussed.

The algorithm starts with ABC optimization with the objective of finding C cluster centers that minimize the inter-cluster fuzzy compactness function. The objective function for the ABC optimization is as equation 7.22.

$$minJ(M, v_1, v_2, ...v_C) = \sum_{i=1}^{C} \sum_{k=1}^{n} (\mu_{ik})^q (d_{ik})^2 \qquad (7.22)$$

where $x_k = kth$ data point (possibly m dimensional vector) and $(k = 1, 2, ...C)$, $v_i =$ the center of the *ith* fuzzy cluster $(i = 1, 2, ...c)$, and $d_{ik} = \|x_k - v_i\|_2 = [\sum_{j=1}^{m} (x_{kj} - v_{ij})^2]^{\frac{1}{2}}$.

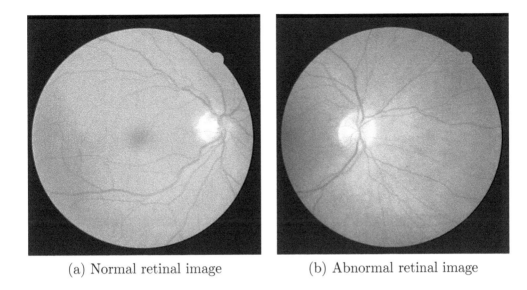

(a) Normal retinal image   (b) Abnormal retinal image

Figure 7.1: DRIVE retinal images: samples for normal and abnormal

The limiting constraints for the variables $v_1, v_2, ..., v_C$ are $\in [0255]$. In ABC searches for C clusters, each represented by the cluster mean, centroid, and hence commonly the search space is of 2D when searching for two clusters, each represented by a single gray level.

Two retinal images databases were used for assessment of the proposed algorithm performance. Both databases are available for free to scientific research on retinal vasculature and they are the most used benchmark databases; namely, DRIVE and STARE. The DRIVE (Digital Retinal Images for Vessel Extraction) [78] is a publicly available database, consisting of a total of 40 color fundus photographs. The photographs were obtained from a diabetic retinopathy screening program in the Netherlands. The screening population consisted of 453 subjects between 31 and 86 years of age. Each image has been JPEG compressed, which is common practice in screening programs. Of the 40 images in the database, 7 contain pathology, namely, exudates, hemorrhages, and pigment epithelium changes.

Figure 7.1 shows examples of normal and abnormal images from DRIVE database. We can remark about the clear confusion between the vessels and the background, especially for images with abnormalities; also we can remark about the dark and white spots, which make the brightness inhomogeneous all over the image.

The second database is the structured analysis of the retina (STARE) originally collected by Hoover et al. [79]. It consists of a total of twenty eye fundus color images where ten of them contain pathology. The images were captured using a TopCon TRV-50 fundus camera with FOV equal to thirty-five degrees. Each image resolution is 700*605

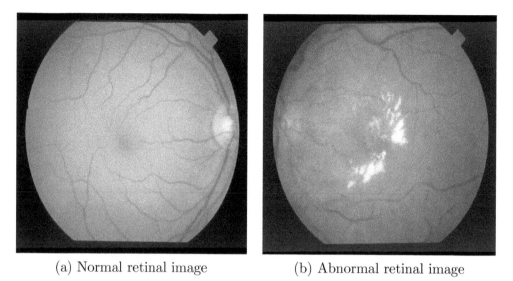

(a) Normal retinal image   (b) Abnormal retinal image

Figure 7.2: STARE retinal images: samples for normal and abnormal

pixels with eight bits per color channel and it is available in PPM format. The set of
twenty images are not divided into separated training and testing sets. The circular FOV
binary mask for each image is not available, so we created it by hand for each image using
MATLAB function, which creates an interactive draggable and resizable polygon on the
image to specify FOV. The database contains two sets of manual segmentations made
by two different ophthalmologists, where the second ophthalmologist's segmentation is
accepted as the ground-truth for performance assessment. Figure 7.2 shows examples of
normal and abnormal images from STARE database. In the STARE database we can
see the same difficulties as the DRIVE dataset, but the STARE contains more abnormal
images.

Three measures are calculated for evaluating the segmentation performance of proposed novel algorithm: sensitivity (Se), specificity (Sp), and accuracy (Acc). These
measures are computed individually for each image and on average for the whole test
images set.

$$Se = TP/(TP + FN) \tag{7.23}$$

$$Sp = TN/(TN + FP) \tag{7.24}$$

$$Acc = (TP + TN)/(TP + FN + TN + FP) \tag{7.25}$$

where the true positives (TP) are the pixels classified as vessel-like and they are really
vessel pixels in the ground-truth. The false negatives (FN) are the pixels classified as
non-vessel although they are really vessel pixels in the ground-truth. The true negatives

(TN) are the pixels classified as non-vessel and they are really background pixels in the ground-truth. The false positives (FP) are the pixels classified as vessel-like although they are really background pixels in the ground-truth. So the sensitivity (Se) is the ratio of correctly classified vessel pixels, while specificity (Sp) is the ratio of correctly classified background pixels, and accuracy (Acc) is the ratio of correctly classified both vessels and background pixels.

Table 7.4 outlines the performance measure values resulting from the proposed system over the DRIVE dataset. ABC is optimized with 100 iterations and population size of 20 and pattern search is initialized with the cluster centers obtained from the ABC. The table outlines the robustness of the algorithm against changes in the input image, where we can remark on the resulted stable accuracy regardless of whether the input image is normal or abnormal, and it gives a fairly same result even with images with exudates, hemorrhages, and pigment epithelium changes. Similar results can be noted in Table 7.3 but with much enhanced accuracy in the STARE database. The beauty of the proposed algorithm is much clearer for the STARE dataset as it contains much abnormal images which the proposed algorithm can handle by the second level optimization.

Figure 7.3 displays sample results for an ABC hybrid with fuzzy C-means cluster where sample image from STARE dataset is displayed and another image from DRIVE dataset is displayed with the corresponding segmentation.

7.4.2 Memetic artificial bee colony for integer programming

Due to the simplicity of the ABC algorithm, it has been applied to solve a large number of problems. ABC is a stochastic algorithm and it generates trial solutions with random moves, however it suffers from slow convergence. In order to accelerate the convergence of the ABC algorithm, we proposed a new hybrid algorithm, which is called memetic artificial bee colony for integer programming (MABCIP). The proposed algorithm is a hybrid algorithm between the ABC algorithm and a random walk with direction exploitation (RWDE) as a local search method. MABCIP is tested on seven benchmark functions. The numerical results demonstrate that MABCIP is an efficient and robust algorithm for solving integer programming problems.

Integer programming problem definition

An integer programming problem is a mathematical optimization problem in which all of the variables are restricted to be integers. The unconstrained integer programming problem can be defined as follow:

$$min f(x), \ \ x \in S \subseteq \mathbb{Z}^n, \tag{7.26}$$

where $\mathbb{Z}$ is the set of integer variables, and S is not necessarily a bounded set.

Table 7.3: The performance measures of the different images in the DRIVE database

Image	Sensitivity	Specificity	Accuracy
1	0.857	0.945	0.933
2	0.798	0.972	0.945
3	0.653	0.983	0.934
4	0.699	0.98	0.942
5	0.661	0.984	0.939
6	0.657	0.978	0.932
7	0.673	0.978	0.937
8	0.63	0.977	0.932
9	0.7	0.971	0.938
10	0.604	0.987	0.94
11	0.734	0.968	0.937
12	0.743	0.968	0.939
13	0.715	0.971	0.934
14	0.781	0.964	0.942
15	0.795	0.961	0.943
16	0.7	0.976	0.939
17	0.644	0.982	0.939
18	0.776	0.962	0.94
19	0.868	0.956	0.945
20	0.774	0.964	0.944
total	0.721	0.971	0.9388

Memetic ABC for integer programming (MABCIP) algorithm

In this section, we highlight the main components of MABCIP algorithm. In the MABCIP algorithm, we tried to combine the ABC algorithm with its ability to global search (exploration process) and the random walk with direction exploitation with its ability to local search (exploitation process). The main steps of the MABCIP algorithm are presented in the following subsection.

Memetic ABC for integer programming

In the memetic ABC for integer programming algorithm, the initial population is generated randomly, which contains NS solutions x_i, $i = \{1, \ldots, NS\}$; each solution is a D dimensional vector. The number of the NS solutions (food sources) is equal to the number of employed bees. The solutions in the initial population are evaluated by calculating their fitness function as follows:

$$f(x_i) = \begin{cases} \frac{1}{1+f(x_i)} & \text{if } f(x_i) \geq 0 \\ 1 + abs(f(x_i)) & \text{if } f(x_i) < 0 \end{cases} \qquad (7.27)$$

Table 7.4: The performance measures of different images in the STARE database

Image	Sensitivity	Specificity	Accuracy
1	0.686	0.952	0.923
2	0.68	0.961	0.935
3	0.802	0.946	0.934
4	0.561	0.984	0.94
5	0.496	0.976	0.916
6	0.736	0.973	0.952
7	0.708	0.993	0.961
8	0.677	0.993	0.961
9	0.692	0.993	0.96
10	0.705	0.963	0.934
11	0.665	0.991	0.959
12	0.735	0.991	0.964
13	0.695	0.989	0.952
14	0.671	0.991	0.95
15	0.712	0.983	0.95
16	0.568	0.989	0.929
17	0.622	0.994	0.948
18	0.48	0.997	0.961
19	0.438	0.994	0.961
20	0.567	0.985	0.946
total	0.649	0.982	0.94677

The best food source is memorized x_{best} and the probability of each food source is calculated in order to generate a new trail solution v_i by onlooker bees. The associated probability of each food source P_i is defined as follows:

$$P_i = \frac{f_i}{\sum_{j=1}^{NS} f_j} \tag{7.28}$$

and the trail solution can be generated as $v_{ij} = x_{ij} + \phi_{ij}(x_{ij} - x_{kj})$, $\phi_{ij} \in [-1, 1]$, $k \in \{1, 2, \ldots, NS\}$, $j \in \{1, 2, \ldots, D\}$, and $i \neq k$. The trail solution is evaluated and if it is better than or equal to the old solution, then the old solution is replaced with the new solutions; otherwise the old solution is retained. The best solution is memorized and the local search algorithm starts to refine the best found solution so far. If the food source cannot improve for a limited number of cycles, which is called *limit*, the food source is considered to be abandoned and replaced with a new food source by scout. The operation is repeated until termination criteria are satisfied, i.e., the algorithm reaches to *MCN* maximum cycle number.

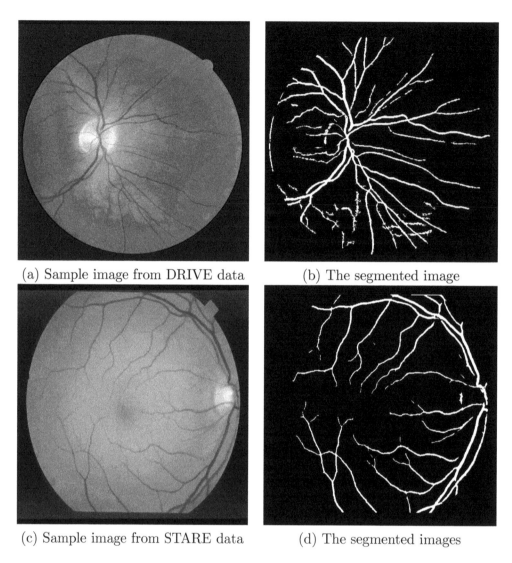

(a) Sample image from DRIVE data   (b) The segmented image

(c) Sample image from STARE data   (d) The segmented images

Figure 7.3: Sample images from DRIVE and STARE database and the corresponding segmented version

Numerical experiments

The efficiency of the MABCIP algorithm is tested on seven benchmark functions [80] and the general performance of it is presented as follows.

The general performance of MABCIP with integer programming functions

The general performance of the proposed algorithm is tested on four benchmark functions (randomly picked) as shown in Figure 7.4. We can observe from Figure 7.4 that the function values for each test function are rapidly converging as the number of iterations increases.

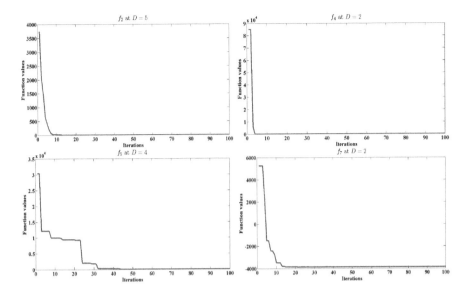

Figure 7.4: The general performance of the MABCIP with integer programming functions

7.5 Chapter conclusion

An overview of the basic concepts of the ABC optimization, its variants, and hybridization with other optimization techniques was discussed. In addition, different real life application areas were discussed, where the ABC optimization algorithm has applied. Moreover, reviews of the results of ABC in retinal vessel segmentation and memetic ABC for integer programming were discussed.

Bibliography

[1] Dervis Karaboga and Bahriye Basturk, "A powerful and efficient algorithm for numerical function optimization: artificial bee colony (ABC) algorithm", *Journal of Global Optimization*, Vol. 39(3), pp. 459-471, 2007.

[2] Jason Brownlee, *Clever Algorithms: Nature-Inspired Programming Recipes*, Jason Brownlee, 2011.

[3] Reza Akbari, Alireza Mohammadi, and Koorush Ziarati, "A novel bee swarm optimization algorithm for numerical function optimization", *Commun Nonlinear Sci Numer Simulat*, pp. 3142-3155, 2010.

[4] Peng-Yeng Yin, "Trends in Developing Metaheuristics, Algorithms, and Optimization Approaches", IGI Global, 2012.

[5] Mustafa Servet Kiran and Mesut Gunduz, "XOR-based artificial bee colony algorithm for binary optimization", *Turkish Journal of Electrical Engineering and Computer Sciences*, Vol. 21, pp. 2307-2328, 2013.

[6] Gaowei Yan and Chuangqin Li, "An effective refinement artificial bee colony optimization algorithm based on chaotic search and application for PID control tuning", *Journal of Computational Information Systems*, Vol. 7(9), pp. 3309-3316, 2011.

[7] Yingsen Hong, Zhenzhou Ji, and Chunlei Liu, "Research of Parallel Artificial Bee Colony Algorithm Based on MPI", in: Proceedings of the 2nd International Conference on Computer Science and Electronics Engineering (ICCSEE 2013), Atlantis Press, Paris, France, pp. 1352-1355.

[8] E. Mezura-Montes and R.E. Velez-Koeppel, "Elitist artificial bee colony for constrained real-parameter optimization", in: 2010 IEEE Congress on Evolutionary Computation (CEC), pp. 1-8, 2010.

[9] Heyin Hakli and Harun Uuz, "Levy flight distribution for scout bee in artificial bee colony algorithm", *Lecture Notes on Software Engineering*, Vol. 1(3), August 2013.

[10] Pei-Wei Tsai, Jeng-Shyang Pan, Bin-Yih Liao, and Shu-Chuan Chu, "Enhanced artificial bee colony optimization", *International Journal of Innovative, Computing, Information and Control* (ICIC2009), Vol. 5(12), pp. 1-08, December 2009.

[11] Ajith Abraham, Ravi Kumar Jatoth, and A. Rajasekhar, "Hybrid differential artificial bee colony algorithm", *Journal of Computational and Theoretical Nanoscience*, Vol. 9, pp. 1-9, 2012.

[12] Noorazliza Sulaiman, Junita Mohamad-Saleh, and Abdul Ghani Abro, "A Modified Artificial Bee Colony (JA-ABC) Optimization Algorithm", in: Proceedings of the 2013 International Conference on Applied Mathematics and Computational Methods in Engineering, pp. 74-79, 2013.

[13] Hai-Bin Duan, Chun-Fang Xu, and Zhi-Hui Xing, "A hybrid artificial bee colony optimization and quantum evolutionary algorithm for continuous optimization problem", *International Journal of Neural Systems*, Vol. 20(1), pp. 39-50, 2010.

[14] J. Li, Q. Pan, S. Xie, and S. Wang, "A hybrid artificial bee colony algorithm for flexible job shop scheduling problems", *International Journal of Computers, Communications and Control*, Vol. VI(2)2, pp. 286-296, June 2011.

[15] Ouz Altun and Tark Korkmaz, Particle Swarm Optimization: "Artificial Bee Colony Chain (PSOABCC): A Hybrid Metahueristic Algorithm", in: International Workshops on Electrical and Computer Engineering Subfields, 22-23 August 2014, Koc University, Istanbul-Turkey, 2014.

[16] Ebru Turanolu, Eren AZceylan, and Mustafa Servet Kiran, "Particle swarm optimization and artificial bee colony approaches to optimize of single input-output

fuzzy membership functions", in: Proceedings of the 41st International Conference on Computers and Industrial Engineering.

[17] Habib Shah, Rozaida Ghazali, Nazri Mohd Nawi, Mustafa Mat Deris, and Tutut Herawan, "Global artificial bee colony-Levenberq-Marquardt (GABCLM) algorithm for classification", *International Journal of Applied Evolutionary Computation*, Vol. 4(3), pp. 58-74, July-September 2013.

[18] Vahid Chahkandi, Mahdi Yaghoobi, and Gelareh Veisi, "Feature selection with chaotic hybrid artificial bee colony algorithm based on fuzzy" (CHABCF), *Journal of Soft Computing and Applications*, pp.1-8, 2013.

[19] Ahcene Habbi and Yassine Boudouaoui, "Hybrid artificial bee colony and least squares method for rule-based systems learning", *World Academy of Science, Engineering and Technology, International Journal of Computer, Control, Quantum and Information Engineering* Vol. 8(12), 2014.

[20] Bin Wu, Cunhua Qian, Weihong Ni, and Shuhai Fan, "Hybrid harmony search and artificial bee colony algorithm for global optimization problems", *Computers and Mathematics with Applications*, Vol. 64, pp. 2621-2634, 2012.

[21] M. Kefayat, A. Lashkar Ara, and S.A. Nabavi Niaki, "A hybrid of ant colony optimization and artificial bee colony algorithm for probabilistic optimal placement and sizing of distributed energy resources", *Energy Conversion and Management*, Vol. 92, pp. 149-161, 2015.

[22] Ahmed Ahmed El-Sawy, Elsayed M. Zaki, and R.M. Rizk-Allah, "Hybridizing ant colony optimization with firefly algorithm for unconstrained optimization problems", *The Online Journal on Computer Science and Information Technology* (OJCSIT), Vol. 3(3), pp. 185-193, 2013.

[23] Mohammad Shokouhifar and Gholamhasan Sajedy Abkenar, "An Artificial Bee Colony Optimization for MRI Fuzzy Segmentation of Brain Tissue", in: International Conference on Management and Artificial Intelligence, IPEDR2011, Vol. 6, IACSIT Press, Bali, Indonesia, 2011.

[24] Adiljan Yimit, Yoshihiro Hagihara, Tasuku Miyoshi, and Yukari Hagihara, "Automatic Image Enhancement by Artificial Bee Colony Algorithm", in: Proc. SPIE, International Conference on Graphic and Image Processing (ICGIP), 2012.

[25] Jaspreet Kaur, Sukwinder Kaur, and Maninder Kaur, "Image enhancement using particle swarm optimization and honey bee", *International Journal of Agriculture Innovations and Research*, Vol. 2(2), pp. 302-304, 2013.

[26] D. Chandrakala and S. Sumathi, "Application of artificial bee colony optimization algorithm for image classification using color and texture feature similarity fusion", *Artificial Intelligence*, Vol. , Article ID 426957, 2012.

[27] Anil Kumar Mishra, Madhabananda Das, and T.C. Panda, "Hybrid swarm intelligence technique for CBIR systems", *IJCSI International Journal of Computer Science Issues*, Vol. 10(2), pp. 1694-0784, March 2013.

[28] Chih-Hsun Lin, Chung-l Huang, Yung-Nien Sun, and Ming-Huwi Horng, "Multimodality Registration by Using Mutual Information with Honey Bee Mating Optimization (HBMO)", in: IEEE EMBS Conference on Biomedical Engineering and Sciences (IECBES), pp. 13-16, 2010.

[29] Yudong Zhang and Lenan Wu, "Rigid Image Registration by PSOSQP Algorithm. Advances in Digital Multimedia", in: *World Science Publisher*, Vol. 1(1), pp. 4-8, United States, March 2012.

[30] Kamalam Balasubramani and Karnan Marcus, "Artificial bee colony algorithm to improve brain MR image segmentation", *International Journal on Computer Science and Engineering* (IJCSE), Vol. 5(1), pp. 31-36, January 2013.

[31] Jzau-Sheng Lin and Shou-Hung Wu, "Fuzzy artificial bee colony system with cooling schedule for the segmentation of medical images by using of spatial information", *Research Journal of Applied Sciences, Engineering and Technology*, Vol. 4(17), pp. 2973-2980, Maxwell Scientific Organization, 2012.

[32] Erik Cuevas, Felipe Sencin, Daniel Zaldivar, "Marco Prez-Cisneros, and Humberto Sossa, A multi-threshold segmentation approach based on artificial bee colony optimization", *Applied Intelligence Journal*, Vol. 37(3), pp. 321-336, October 2012.

[33] Sushil Kumar, Tarun Kumar Sharma, Millie Pant, and A.K. Ray, "Adaptive Artificial Bee Colony for Segmentation of CT Lung Images", in: International Conference on Recent Advances and Future Trends in Information Technology (iRAFIT), Proceedings published in International Journal of Computer Applications (IJCA), 2012.

[34] M.M. George, M. Karnan, and R. SivaKumar, "Supervised artificial bee colony system for tumor segmentation in CT/MRI images", *International Journal of Computer Science and Management Research*, Vol. 2(5), pp. 2529-2533, May 2013.

[35] Tahir Sag and Mehmet Cunkas, "Development of image segmentation techniques using SWARM intelligence", *ICCIT*, 2012.

[36] Yudong Zhang and Lenan Wu, "Optimal multi-level thresholding based on maximum Tsallis entropy via an artificial bee colony approach", *Entropy*, Vol. 13(4), pp. 841-859, 2011.

[37] Miao Maa, Jianhui Liang, Min Guo, Yi Fan, and Yilong Yin, "SAR image segmentation based on artificial bee colony algorithm", *Applied Soft Computing*, Vol. 11, pp. 5205-5214, 2011.

[38] Y. Zhang, L. Wu, and S. Wang, "Magnetic resonance brain image classification by an improved artificial bee colony algorithm", *Progress in Electromagnetics Research*, Vol. 116, pp. 65-79, 2011.

[39] B. Akay, and D. Karaboga, Wavelet Packets Optimization Using Artificial Bee Colony Algorithm, in: 2011 IEEE Congress on Evolutionary Computation (CEC), pp. 89-94, 2011.

[40] M.H. Horng, and T.W. Jiang, "Multilevel Image Thresholding Selection Using the Artificial Bee Colony Algorithm", in: Wang F, Deng H, Gao Y, Lei J (eds) *Artificial Intelligence and Computational Intelligence*. Lecture Notes in Computer Science, Vol. 6320, Springer, Berlin, pp. 318-325, 2011.

[41] M.H, Horng, "Multilevel thresholding selection based on the artificial bee colony algorithm for image segmentation", *Expert Syst Appl*, 38(11), pp. 13785-13791, 2011.

[42] C. Xu, and H. Duan, "Artificial bee colony (abc) optimized edge potential function (epf) approach to target recognition for low-altitude aircraft". *Pattern Recognition Letter*, Vol. 31(13, SI), pp. 1759-1772, 2011.

[43] Prabhat Kumar Sharma, V.S. Bhavya, K.M. Navyashree, K.S. Sunil, and P. Pavithra, "Artificial bee colony and its application for image fusion", *International Journal Information Technology and Computer Science*, Vol. 11, pp. 42-49, 2012.

[44] Prabhat Kumar Sharma, V.S. Bhavya, K.M. Navyashree, K.S. Sunil, and P. Pavithra, "Artificial bee colony and its application for image fusion". *International Journal of Information Technology and Computer Science*, Vol. 11, pp. 42-49, 2012.

[45] C. Chidambaram, and H.S. Lopes, "A New Approach for Template Matching in Digital Images Using an Artificial Bee Colony Algorithm", in: Abraham A, Herrera F, Carvalho A, Pai V (eds) 2009 World Congress on Nature and Biologically Inspired Computing (NABIC 2009), pp. 146-151, 2009.

[46] S. Nebti, A. Boukerram, "Handwritten Digits Recognition Based on Swarm Optimization Methods, in: Zavoral F, Yaghob J, Pichappan P, El-Qawasmeh E (eds) Networked Digital Technologies". *Communications in Computer and Information Science*, Vol. 87, Springer, Berlin, pp. 45-54, 2010.

[47] Ashita S. Bhagade and Parag V. Puranik, "Artificial bee colony (ABC) algorithm for vehicle routing optimization problem", *International Journal of Soft Computing and Engineering* (IJSCE), Vol. 2(2), May 2012.

[48] W.Y. Szeto, Y. Wu, and S.C. Ho, "An artificial bee colony algorithm for the capacitated vehicle routing problem", *European Journal of Operational Research*, Vol. 215(1), pp. 126-135, 2011.

[49] B. Yao, C. Yang, J. Hu, and B. Yu, "The optimization of urban subway routes based on artificial bee colony algorithm", in: Chen F, Gao L, Bai Y (eds) *Key Technologies of Railway Engineering High Speed Railway, Heavy Haul Railway and Urban Rail Transit*, Beijing Jiaotong University, Beijing, pp. 747-751, 2010.

[50] Plamenka Borovska, Veska Gancheva, and Nikolay Landzhev, "Code Optimization and Scalability Testing of an Artificial Bee Colony Based Software for Massively Parallel Multiple Sequence Alignment on the Intel MIC Architecture", www.prace-ri.eu, 2014.

[51] Y. Zhang, L. Wu, and S. Wang, "Magnetic resonance brain image classification by an improved artificial bee colony algorithm", *Progress in Electromagnetics Research*, Vol. 116, pp. 65-79, 2011.

[52] Janaki Sathya and K. Geetha, "Mass classification in breast DCE-MR images using an artificial neural network trained via a bee colony optimization algorithm", *Science Asia*, Vol. 39, pp. 294-306, 2013.

[53] DuAn Teodorovi, Milica Elmi, and Ljiljana Mijatovi-Teodorovi, "Combining case-based reasoning with bee colony optimization for dose planning in well differentiated thyroid cancer treatment", *Journal Expert Systems with Applications, An International Journal archive*, Vol. 40(6), pp. 2147-2155, May 2013.

[54] M.S.P. Babu, N.T., Rao, "Implementation of artificial bee colony (abc) algorithm on garlic expert advisory system", *International Journal of Computer Science Research*, Vol. 1(1), pp. 69-74, 2007.

[55] Dervis Karaboga and Bahriye Basturk, "A powerful and efficient algorithm for numerical function optimization: artificial bee colony" (ABC) algorithm, *Journal of Global Optimization*, Vol. 39(3), pp. 459-471, 2007.

[56] D. Karaboga, C. Ozturk, "Neural networks training by artificial bee colony algorithm on pattern classification", *Neural Network World*, Vol. 19(3), pp. 279-292 2009.

[57] S.N. Omkar and J. Senthilnath, "Artificial bee colony for classification of acoustic emission signal source", *Int J Aerosp Innov*, Vol. 1(3), pp. 129-143, 2009.

[58] R. Irani and R. Nasimi, "Application of artificial bee colony-based neural network in bottom hole pressure prediction in under balanced drilling", *J Pet Sci Eng*, Vol. 78(1), pp. 6-12, 2011.

[59] T. Kurban and E. Besdok, "A comparison of RBF neural network training algorithms for inertial sensor based terrain classification", *Sensors*, Vol. 9(8), pp. 6312-6329, 2009.

[60] T.J. Hsieh and W.C. Yeh, "Knowledge discovery employing grid scheme least squares support vector machines based on orthogonal design bee colony algorithm", IEEE Trans Syst Man Cybern, Part B: Cybern, Vol. 41(5), pp. 1198-1212, 2011.

[61] P.Y. Kumbhar and S. Krishnan, "Use of artificial bee colony (abc) algorithm in artificial neural network synthesis", *Int J Adv Eng Sci Technol*, Vol. 11(1), pp. 162-171, 2011.

[62] N. Suguna and K. Thanushkodi, "A novel rough set reduct algorithm for medical domain based on bee colony optimization", *Journal of Computing*, Vol. 2(6), pp. 49-54, June 2010.

[63] H. Li, J. Li, and F. Kang, "Risk analysis of dam based on artificial bee colony algorithm with fuzzy c-means clustering", *Can J Civ Eng*, Vol. 38(5), pp. 483-492, 2011.

[64] I. Eke, M.C. Taplamacioglu, and I. Kocaarslan, "Design of robust power system stabilizer based on artificial bee colony algorithm", *J Fac Eng Arch Gazi Univ*, Vol. 26(3), pp. 683-690, 2011.

[65] N. Suguna and K.G. Thanushkodi, "An independent rough set approach hybrid with artificial bee colony algorithm for dimensionality reduction", *Am J. Appl. Sci*, Vol. 8(3), pp. 261-266, 2011.

[66] M.A.M. Shukran, Y.Y. Chung, W.C. Yeh, N. Wahid, and A.M.A. Zaidi, "Artificial bee colony based data mining algorithms for classification tasks", *Mod Appl Sci*, Vol. 5(4), pp. 217-231, 2011.

[67] C. Zhang, D. Ouyang, and J. Ning, "An artificial bee colony approach for clustering", *Expert Syst Appl*, Vol. 37(7), pp. 4761-4767, 2010.

[68] R.V. Rao and P.J. Pawar, "Modelling and optimization of process parameters of wire electrical discharge machining", *Proc Inst Mech Eng Part B-J Eng Manuf*, Vol. 223(11), pp. 1431-1440, 2009.

[69] A. Gomez-Iglesias, M.A. Vega-Rodriguez, F. Castejon, M. Cardenas-Montes, and E. Morales-Ramos, "Artificial Bee Colony Inspired Algorithm Applied to Fusion Research in a Grid Computing Environment", in: 18th Euromicro International Conference on Parallel, Distributed and Network-Based Processing (PDP), pp. 508-512, 2010.

[70] A. Hadidi, S.K. Azad, and S.K. Azad, "Structural Optimization Using Artificial Bee Colony Algorithm", in: 2nd International Conference on Engineering Optimization, 2010.

[71] E. Bijami, M. Shahriari-kahkeshi, and H. Zamzam, "Simultaneous Coordinated Tuning of Power System Stabilizers Using Artificial Bee Colony Algorithm", in: 26th International Power System Conference (PSC), pp. 1-8, 2011.

[72] A.K. Sarma and K.M. Rafi, "Optimal capacitor placement in radial distribution systems using artificial bee colony (abc) algorithm", *Innov Syst Des Eng*, Vol. 2(4), pp. 177-185, 2011.

[73] H. Gozde and M.C. Taplamacioglu, "Comparative performance analysis of artificial bee colony algorithm for automatic voltage regulator (AVR) system", *J Frankl Inst-Eng Appl Math*, Vol. 348(8), pp. 1927-1946, 2011.

[74] S.K. Udgata, S.L. Sabat, and S. Mini, "Sensor Deployment in Irregular Terrain Using Artificial Bee Colony Algorithm", in: Abraham A, Herrera F, Carvalho A, Pai V (ed.) 2009 World Congress on Nature and Biologically Inspired Computing (NABIC 2009), pp. 1308-1313, 2009.

[75] S.K. Udgata, S.L. Sabat, and S. Mini, "Sensor Deployment in Irregular Terrain Using Artificial Bee Colony Algorithm", in: Abraham A, Herrera F, Carvalho A, Pai V (ed.) 2009 World Congress on Nature and Biologically Inspired Computing (NABIC 2009), pp. 1308-1313.

[76] C. Ozturk and D. Karaboga, "Hybrid Artificial Bee Colony Algorithm for Neural Network Training", in: 2011 IEEE Congress on Evolutionary Computation (CEC), pp. 84-88, 2011.

[77] C. Ztrk, D. Karaboga, and B. Grkemli, "Artificial bee colony algorithm for dynamic deployment of wireless sensor networks", *Turk J Electr Eng Comput Sci*, Vol. 20(2), pp. 1-8, 2012.

[78] J.J. Staal, M.D. Abramoff, M. Niemeijer, M.A. Viergever, and B.V. Ginneken, "Ridge based vessel segmentation in color images of the retina", *IEEE Trans. Med. Imaging*, Vol. 23(4), pp. 501-509, 2004.

[79] A. Hoover and M. Goldbaum, "Locating the optic nerve in a retinal image using the fuzzy convergence of the blood vessels", IEEE Transactions on Medical Imaging, vol. 22 no. 8, pp. 951-958, August 2003.

[80] K.E. Parsopoulos and M.N. Vrahatis, "Unified Particle Swarm Optimization for Tackling Operations Research Problems", in: Proceedings of IEEE 2005 Swarm Intelligence Symposium, Pasadena, Calif., USA, pp. 53-59, 2005.

8 Wolf-Based Search Algorithms

8.1 Wolf search algorithm (WSA)

WSA is a new bio-inspired heuristic optimization algorithm that imitates the way wolves search for food and survive by avoiding their enemies. The WSA was proposed by Rui Tang et al. [1].

8.1.1 Wolves in nature

Wolves are social predators that hunt in packs. Wolves typically commute as a nuclear family. They remain silent and use stealth when hunting prey together and have developed unique, semi-cooperative characteristics; that is, they move in a group in a loosely coupled formation, but tend to take down prey individually [1]. WSA naturally balances scouting the problem space in random groups (breadth) and searching for the solution individually (depth).

When hunting, wolves simultaneously search for prey and watch out for threats such as human hunters or tigers. Each wolf in the pack chooses its own position, continuously moving to a better spot and watching for potential threats [1].

WSA is equipped with a threat probability that simulates incidents of wolves bumping into their enemies. When this happens, the wolf dashes a great distance away from its current position, which helps break the deadlock of getting stuck in local optima. The direction and distance they travel when moving away from a threat are random.

Each wolf in the WSA has a sensing distance that creates a sensing radius or coverage area generally referred to as visual distance. This visual distance is applied to the search for food (the global optimum), an awareness of their peers (in the hope of moving into a better position), and signs that enemies might be nearby (for jumping out of visual range). Once they sense that prey is near, they approach quickly, quietly, and very cautiously because they do not want to be discovered.

In search mode, when none of the above-mentioned items, threat or prey, are detected within visual range, the wolves move in Brownian motion (BM). Artificial wolf search main steps are given in the following algorithm.

8.1.2 Artificial wolf search algorithm

Three rules can be formulated to describe the wolf searching that can be summarized as follows:

- Each wolf has a fixed visual area with a radius. In hyperplane, where multiple attributes dominate, the distance would be estimated by Minkowski distance; see equation 8.1.

$$d(x_i, x_c) = (\sum_{k=1}^{n} |x_{i,k} - x_{c,k}|^\lambda)^{\frac{1}{\lambda}} \tag{8.1}$$

  where x_i is the current position, x_c are all the potential neighboring positions near x_i, and λ is commonly 1 or 2.

- The fitness of the objective function represents the quality of the wolf's current position. The wolf always tries to move to a better position but rather than choose the best terrain it opts to move to a better position that already houses a companion. If there is more than one better position, the wolf will choose the best terrain inhabited by another wolf from the given options. Otherwise, the wolf will continue to move randomly in BM.

- It is possible that the wolf will sense an enemy. The wolf will escape to a random position far from the threat and beyond its visual range.

WSA features three different types of preying behavior that take place in sequence.

1. *Preying initiatively*: This step allows the wolf to check its visual perimeter to detect prey. Once the prey is found within the wolf's visual distance, it will move toward the prey that has the highest or fitness value; see equation 8.2, in which circumstance the wolf will omit looking out for its companions. In WSA, this is reflected by the fact that the wolf will change its own position for that of the prey, which has the highest value, and because no other position is higher than the highest, the wolf will maintain this direction.

$$x_i^{t+1} = x_i^t + \beta_0 * \exp^{-r^2}(x_j^t - x_i^t) + escape \tag{8.2}$$

  where r is the distance between the wolf and its peer with the better location; escape() is a function that calculates a random position to jump to with a constraint of minimum length; v, x is the wolf, which represents a candidate solution; and x_j^t is the peer with a better position as represented by the value of the fitness function.

2. *Preying passively*: If the wolf does not find any food or better shelter inhabited by a peer in the previous step, then it will prey passively. In this passive mode, the wolf only stays alert for incoming threats and attempts to improve its current position by comparing it to those of its peers.

$$x_i^{t+1} = x_i^t + \alpha * r * rand \tag{8.3}$$

input : r = Radius of the visual range.

s = Step size by which a wolf changes its position at a time

α = Velocity factor of wolf

p_a = A user-defined threshold $[0..1]$, determines how frequently an enemy appears.

n Number of wolves.

$Iter_{max}$ maximum allowed number of iterations.

output: Optimal Wolf's position and its fitness

Generate initial population of n random positions

;

while *Stopping criteria not met* **do**

> **foreach** $Wolf_i$ **do**
>
> Calculate $wolf_i$ new position using BM motion; see equation 8.3.
>
> If The wolf's calculated new position is better than its current position Update the wolf's current position.
>
> Find Wolves in the visual domain of $Wolf_i$
>
> **if** *none of the wolf's exist* **then**
>
> > Calculate $wolf_i$ new position using BM motion; see equation 8.3.
> >
> > If The wolf's calculated new position is better than its current position Update the wolf's current position.
>
> **else**
>
> > Calculate new position by moving towards the best wolf in the visual domain; see equation 8.4 If The wolf's calculated new position is better than its current position Update the wolf's current position.
>
> **end**
>
> IFrand $> p_a$ Escape to new position; see equation 8.4 Update the best solution

end

Algorithm 9: Wolf search algorithm

where α and r are constants and *rand* is a random number drawn from uniform distribution in the range $[0,1]$.

3. *Escape*: When a threat is detected, given a threat probability p_a, the wolf escapes very quickly by relocating itself to a new position with an escape distance that is greater than its visual range. The emergence of threats is modeled randomly at a probability defined by the user. Escape is an important step that helps keep all of the wolves from falling into and getting stuck at a local optimum.

$$x_i^{t+1} = x_i^t + \alpha. \, s. \, escape() \tag{8.4}$$

where x_i^t is the wolfs location, α is the velocity, s is the step size, and *escape* is a custom function that randomly generates a position greater than the visual distance v and less than half of the solution boundary.

8.1.3 Wolf search algorithm variants

Many algorithms are proposed in literature inspired from the wolf's search manner, which will be discussed in the coming subsections.

8.1.4 Wolf pack algorithm (WPA)

Wolves are social animals that live in packs. Always there is a *lead wolf*; some elite wolves act as *scouts* and some *ferocious* wolves in a wolf pack [2]. First, the *lead* wolf, is always the smartest and most ferocious one. It is responsible for *commanding* the wolves and *making decisions* by evaluating the surrounding situation and perceiving information from other wolves.

Second, the *lead* wolf sends some elite wolves to hunt around and look for prey in the probable scope. Those elite wolves are *scouts*. They walk around and independently make decisions according to the concentration of smell left by prey; higher concentration means the prey is closer to the wolves.

Third, once a scout wolf finds the trace of prey, it will howl and report that to the lead wolf. Then the lead wolf will evaluate this situation and make a decision whether to summon the ferocious wolves to round up the prey. If they are summoned, the ferocious wolves will move fast toward the direction of that scout wolf.

Fourth, after capturing the prey, the prey is not distributed equitably, but in an order from the strong to the weak.

The predation behavior of the wolf pack is abstracted in three intelligent behaviors, *scouting*, *calling*, and *besieging* behavior, and two intelligent rules, *winner-take-all* generating rule for the lead wolf and the *stronger-survive* renewing rule for the wolf pack [2].

Winner-take-all: The wolf with the best objective function value is selected in each iteration as the lead wolf.

Scouting behavior: Some of the elite wolves except the lead wolf are considered as the scout wolves; they search the solution in predatory space. The state of the scout wolf i is formulated below.

$$x_{id}^p = x_{id} + sin(2\pi * \frac{p}{h}) * step_a^d \tag{8.5}$$

where $step_a$ is suitable step length for wolf in the scouting state, p is the dimension number, and h is a random integer between 1 and p.

Calling behavior: The lead wolf will howl and summon ferocious wolves to gather around the prey. Here, the position of the lead wolf is considered as the one of the prey so that the ferocious wolves aggregate toward the position of lead wolf. $step_b$ is the step length; g_d^k is the position of artificial lead wolf in the dth variable space at the kth

iteration. The position of the ferocious wolf i in the kth iterative calculation is updated according to the following equation:

$$x_{id}^{k+1} = x_{id}^k + step_b^d \cdot \frac{g_d^k - x_{id}^k}{|g_d^k - x_{id}^k|} \tag{8.6}$$

where $step_b$ is the step length and g_d^k is the position of the lead wolf.

Besieging behavior: After large steps running toward the lead wolf, the wolves are close to the prey, then all wolves except the lead wolf will take besieging behavior for capturing prey. Now, the position of lead wolf is considered as the position of prey. g_d^k represents the position of prey; best, in the dth variable space at the kth iteration. The position of wolf i is updated according to the following equation:

$$x_{id}^{k+1} = x_{id}^k + \lambda.step_c^d |g_d^k - x_{id}^k| \tag{8.7}$$

where λ is a random number drawn from uniform distribution in the interval $[-11]$, $step_c$ is the step length for wolf i in the besieging behavior.

The *stronger-survive* rule for the wolf pack: The prey is distributed from the strong to the weak, which will result in some weak wolves dead. The algorithm will generate R wolves while deleting R wolves with bad objective function values. Specifically, with the help of the lead wolf's hunting experience, in the dth variable space, position of the ith one of R wolves is defined as follows:

$$x_{id}^{k+1} = g_d.rand \tag{8.8}$$

g_d is the global best position, i is wolf index in the set $1, 2, ..., R$, and $rand$ is a random number drawn from uniform distribution in the range $[-ss]$ and s is a small number. The flow chart describing the complete WPA algorithm is outlined in Figure 8.1.

8.1.5 Gray wolf optimization (GWO)

Gray wolves are considered as apex predators, meaning that they are at the top of the food chain. Gray wolves mostly prefer to live in a pack. The pack size is 512 on average. They have a very strict social dominant hierarchy.

The leaders are a male and a female, called alphas. The alpha is mostly responsible for making decisions about hunting, sleeping place, time to wake, and so on. The alpha's decisions are dictated to the pack [3].

The second level in the hierarchy of gray wolves is beta. The betas are subordinate wolves that help the alpha in decision-making or other pack activities. The beta wolf can be either male or female, and he/she is probably the best candidate to be the alpha in case one of the alpha wolves passes away or becomes very old.

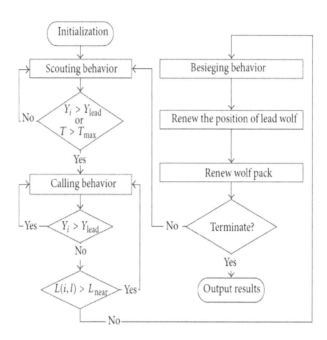

Figure 8.1: The wolf pack optimization algorithm flow chart (after [2])

The lowest ranking gray wolf is omega. The omega plays the role of scapegoat. Omega wolves always have to submit to all the other dominant wolves. They are the last wolves that are allowed to eat.

The fourth class is called subordinate (or delta in some references). Delta wolves have to submit to alphas and betas, but they dominate the omega. *Scouts, sentinels, elders, hunters,* and *caretakers* belong to this category. *Scouts* are responsible for watching the boundaries of the territory and warning the pack in case of any danger. *Sentinels* protect and guarantee the safety of the pack. *Elders* are the experienced wolves who used to be alpha or beta. *Hunters* help the alphas and betas when hunting prey and providing food for the pack. Finally, the *caretakers* are responsible for caring for the weak, ill, and wounded wolves in the pack.

In the mathematical model for the GWO the fittest solution is called the alpha (α). The second and third best solutions are named beta (β) and delta (δ), respectively. The rest of the candidate solutions are assumed to be omega (ω). The hunting is guided by α, β, and δ and the ω follows these three candidates.

In order for the pack to hunt a prey they first encircl: it. In order to mathematically model encircling behavior the following equations are used:

$$\vec{X}(t+1) = \vec{X}_p(t) + \vec{A}.\vec{D} \qquad (8.9)$$

where $\vec{D}$ is as defined in equation 8.10 and t is the iteration number, $\vec{A}$, $\vec{C}$ are coefficient vectors, $\vec{X}_p$ is the prey position, and $\vec{X}$ is the gray wolf position.

$$\vec{D} = |\vec{C}.\vec{X}_p(t) - \vec{X}(t)| \tag{8.10}$$

The $\vec{A}$, $\vec{C}$ vectors are calculated as in equations 8.11 and 8.12.

$$\vec{A} = 2\vec{A}.\vec{r_1} - \vec{a} \tag{8.11}$$

$$\vec{C} = 2\vec{r_2} \tag{8.12}$$

where components of $\vec{a}$ are linearly decreased from 2 to 0 over the course of iterations and r_1, r_2 are random vectors in $[0, 1]$. The hunt is usually guided by the alpha. The beta and delta might also participate in hunting occasionally. In order to mathematically simulate the hunting behavior of gray wolves, the alpha (best candidate solution), beta, and delta are assumed to have better knowledge about the potential location of prey. The first three are the best solutions obtained so far and oblige the other search agents (including the omegas) to update their positions according to the position of the best search agents. So the updating for the wolves' positions is as in equations 8.13, 8.14, and 8.15.

$$\vec{D_\alpha} = |\vec{C_1}.\vec{X_\alpha} - \vec{X}|, \vec{D_\beta} = |\vec{C_2}.\vec{X_\beta} - \vec{X}|, \vec{D_\delta} = |\vec{C_3}.\vec{X_\delta} - \vec{X}| \tag{8.13}$$

$$\vec{X_1} = |\vec{X_\alpha} - \vec{A_1}.\vec{D_\alpha}|, \vec{X_2} = |\vec{X_\beta} - \vec{A_2}.\vec{D_\beta}|, \vec{X_3} = |\vec{X_\delta} - \vec{A_3}.\vec{D_\delta}| \tag{8.14}$$

$$\vec{X}(t+1) = \frac{\vec{X_1} + \vec{X_2} + \vec{X_3}}{3} \tag{8.15}$$

A final note about the GWO is the updating of the parameter $\vec{a}$ that controls the trade-off between exploration and exploitation. The parameter $\vec{a}$ is linearly updated in each iteration to range from 2 to 0 according to equation 8.16.

$$\vec{a} = 2 - t\frac{2}{MaxIter} \tag{8.16}$$

where t is the iteration number and $MaxIter$ is the total number of iterations allowed for the optimization.

8.2 Wolf search optimizers in real world applications

Gray wolf optimizer was used in [4] to find optimal feature subset in order to maximize the classification performance. The fitness function is used is the classification performance at a given selection of feature subset. The used number of variables is the same as the feature vector size of the given data where individual variable value ranges from 0 to 1 and the feature corresponding to a given variable is selected if it passes a given threshold. Different initialization strategies are used to assess the dependence of GWO on the initial solutions used.

Wolf search algorithm was employed in [6] in the domain of feature selection in combination with rough-set to find optimal feature subset to maximize rough-set classifier performance. In that work, authors made use of the fitness function in equation 8.17.

$$\downarrow Fitness = \alpha\gamma(D) + \beta\frac{\mid C - R \mid}{\mid C \mid} \tag{8.17}$$

where $\gamma(D)$ is the classification quality for the selected features in D on the validation set, C is the total number of attributes, R is the number of features selected, and α, β are selected constants to balance the trade-off between accuracy and selected feature size. The used fitness function incorporates both classification accuracy and reduction size.

In [5] gray wolf optimizer has been employed to adjusting PID controller parameters in DC motors. The fitness function to be minimized is as follows:

$$fitness = meanf^p_{init}(Ra, k) \tag{8.18}$$

where K is the number of parameters and Ra is electrical resistance. The fitness function depends on electrical resistance changes and K parameter and develops the output velocity of DC motor.

8.2.1 Gray wolf optimization for feature reduction

In this experiment, a multi-objective version of gray wolf optimization has been used in the feature selection problem. A fitness function was used in this study to exploit the data characterization capabilities of filter-based feature selection methods and the classification performance of wrapper-based feature selection method. The proposed fitness function made use of mutual information index as a guide to direct the search toward finding a feature combination with minor redundancy and maximal relation to class labels. The used mutual information fitness function was formulated as

$$\theta = V - P \tag{8.19}$$

where V is the average mutual information between the selected features and the class labels and P is the average mutual information among the selected features, where V and P are calculated as

$$V = \frac{1}{n}\sum_{i=1}^{n} I(X_i, Y) \tag{8.20}$$

$$P = \frac{1}{n^2}\sum_{i=1}^{n}\sum_{j=1}^{n} I(X_i, X_j) \tag{8.21}$$

where $I(X_i, Y)$ is the mutual information between feature i and the class labels Y and $I(X_i, X_j)$ is the mutual information between feature i and feature j and is defined by

Table 8.1: Description of the data sets used in experiments

Dataset	No. of Features	No. of Samples
Lymphography	18	148
Zoo	16	101
Vote	16	300
Breastcancer	9	699
M-of-N	13	1000
Exactly	13	1000
Exactly2	13	1000
Tic-tac-toe	9	958

Table 8.2: Parameter setting for gray wolf optimization

Parameter	Value(s)
No. of wolves	5
No. of iterations	100
Problem dimension	Same as number of features in any given database
Search domain	[0 1]

the following formulas:

$$I(X_i, Y) = H(X_i) + H(Y) - H(X_i, Y) \tag{8.22}$$

$$I(X_i, X_j) = H(X_i) + H(X_j) - H(X_i, X_j) \tag{8.23}$$

The used fitness function represents the predictability of features from each other and the predictability between individual features. Hence the goodness of a feature combination is estimated as how much the selected features can correctly predict the output class labels and how much they are independent. The range for the parameter $\vec{a}$ that controls the trade-off between exploitation and exploration is limited to the range from 2 to 1 rather than from 2 to 0 to keep solution diversity and to tolerate stagnation.

At the second level optimization, GWO uses wrapper-based principles to further enhance the selected features. The classification performance is used as an objective for the optimization as follows:

$$Fitness = CCR(D) \tag{8.24}$$

where $CCR(D)$ is the correct classification ratio at feature set D.

The optimization in this second phase is much guided toward enhancing the classification accuracy given a preselected classifier: K-nearest neighbor in the current case, but the individual evaluation is more time consuming than the one used in the first stage. Thus, the first stage is used to motivate the search agents to regions with expected promising regions in the feature space, while the second level optimization uses

Table 8.3: Result of different runs for GA, PSO, and GWO of fitness function

Dataset	Breastcancer								
	KNN			MI			MI and KNN		
Algorithm	GA	PSO	GWO	GA	PSO	GWO	GA	PSO	GWO
Mean fitness	0.024	0.030	0.026	-0.378	0.916	-0.412	0.024	0.028	0.027
Std fitness	0.007	0.009	0.006	0.066	0.129	0.032	0.007	0.008	0.007
Best fitness	0.017	0.021	0.021	-0.426	-0.426	-0.426	0.017	0.021	0.021
Worst fitness	0.034	0.043	0.034	-0.274	-0.130	-0.355	0.034	0.043	0.039

Dataset	Exactly								
	KNN			MI			MI and KNN		
Algorithm	GA	PSO	GWO	GA	PSO	GWO	GA	PSO	GWO
Mean fitness	0.269	0.297	0.188	0.003	0.003	0.003	0.292	0.309	0.101
Std fitness	0.051	0.017	0.142	0.000	0.000	0.000	0.011	0.020	0.120
Best fitness	0.180	0.278	0.018	0.003	0.003	0.003	0.275	0.290	0.015
Worst fitness	0.305	0.320	0.305	0.004	0.004	0.004	0.302	0.338	0.302

Dataset	Exactly2								
	KNN			MI			MI and KNN		
Algorithm	GA	PSO	GWO	GA	PSO	GWO	GA	PSO	GWO
Mean fitness	0.233	0.246	0.235	0.003	0.004	0.003	0.240	0.241	0.233
Std fitness	0.012	0.009	0.010	0.000	0.000	0.000	0.015	0.012	0.012
Best fitness	0.219	0.236	0.225	0.003	0.003	0.003	0.228	0.233	0.219
Worst fitness	0.249	0.260	0.248	0.004	0.004	0.003	0.263	0.263	0.249

Dataset	Lymphography								
	KNN			MI			MI and KNN		
Algorithm	GA	PSO	GWO	GA	PSO	GWO	GA	PSO	GWO
Mean fitness	0.167	0.152	0.127	-0.100	-0.092	-0.194	0.148	0.180	0.132
Atd fitness	0.039	0.034	0.051	0.007	0.009	0	0.071	0.034	0.023
Best fitness	0.122	0.102	0.061	-0.111	-0.103	-0.194	0.061	0.143	0.102
Worst fitness	0.224	0.184	0.204	-0.093	-0.084	-0.194	0.245	0.224	0.163

exploitation to intensively find the solution with best classification performance. The parameter $\vec{a}$ used by the GWO to control the diversification and intensification is set in this second level of optimization to the range from 1 to 0 to enhance the intensification of the solutions. This parameter choice allows for less deviation from the initial solutions to this second stage of optimization and allows also for fine tuning to find classification-performance guided solutions.

Table 8.1 summarizes the eight used datasets for further experiments. The datasets are drawn from the UCI data repository [?]. The data are divided into three equal parts: one for *training*, the second part for *validation*, and the third part for *testing*.

Table 8.3: Result of different runs for GA, PSO, and GWO of fitness function (continued)

Dataset	M-of-N								
	KNN			MI			MI and KNN		
Algorithm	GA	PSO	GWO	GA	PSO	GWO	GA	PSO	GWO
Mean fitness	0.097	0.125	0.074	-0.048	-0.0491	-0.053	0.109	0.585	0.028
Std fitness	0.041	0.037	0.064	0.006	0.005	0.000	0.038	0.085	0.011
Sest fitness	0.036	0.087	0.018	-0.053	-0.053	-0.053	0.066	0.030	0.018
Worst fitness	0.150	0.177	0.146	-0.041	-0.040	-0.0530	0.165	0.036	0.042
Dataset	Tic-tac-toe								
	KNN			MI			MI and KNN		
Algorithm	GA	PSO	GWO	GA	PSO	GWO	GA	PSO	GWO
Mean fitness	0.226	0.25	0.236	-0.018	-0.011	-0.018	0.245	0.242	0.221
Std fitness	0.0213	0.013	0.029	0	0.004	0	0.0198	0.028	0.017
Best fitness	0.203	0.231	0.203	-0.018	-0.018	-0.018	0.219	0.203	0.203
Worst fitness	0.253	0.269	0.266	-0.018	-0.008	-0.018	0.269	0.272	0.241
Dataset	Vote								
	KNN			MI			MI and KNN		
Algorithm	GA	PSO	GWO	GA	PSO	GWO	GA	PSO	GWO
Mean fitness	0.054	0.056	0.054	-0.209	-0.198	-0.444	0.06	0.056	0.058
Std fitness	0.0114	0.009	0.021	0.136	0.144	0	0.023	0.019	0.019
Best fitness	0.04	0.04	0.03	-0.444	-0.444	-0.444	0.04	0.04	0.04
Worst fitness	0.07	0.06	0.08	-0.103	-0.094	-0.444	0.1	0.09	0.09
Dataset	Zoo								
	KNN			MI			MI and KNN		
Algorithm	GA	PSO	GWO	GA	PSO	GWO	GA	PSO	GWO
Mean fitness	0.076	0.076	0.076	-0.310	-0.280	-0.529	0.094	0.112	0.083
std fitness	0.049	0.049	0.053	0.142	0.025	0.078	0.024	0.052	0.043
best fitness	0	0	0	-0.565	-0.308	-0.564	0.061	0.031	0.030
Worst fitness	0.118	0.118	0.147	-0.224	-0.244	-0.389	0.118	0.176	0.147

GWO algorithm is compared with the particle swarm optimization (PSO) [53] and genetic algorithms (GA) [52] which are common for space searching. The parameter set for the GWO algorithm is outlined in Table 8.2. The same number of agents and same number of iterations are used for GA and PSO.

Tables 8.3 summarizes the result of running the different optimization algorithms for 10 different runs. Mean fitness function obtained by the gray wolf optimizer achieves remarkable advance over PSO and GA using the different fitness functions over the different datasets used, which ensures the searching capability of GWO. By remarking standard deviation of the solution obtained on the different runs of individual algorithms

Table 8.4: Experiment results of mean classification accuracy

Dataset	KNN			MI			MI and KNN		
	GA	PSO	GWO	GA	PSO	GWO	GA	PSO	GWO
Breastcancer	0.953	0.959	0.947	0.887	0.916	0.924	0.949	0.960	0.960
Exactly	0.726	0.673	0.805	0.664	0.671	0.676	0.685	0.669	0.910
Exactly2	0.748	0.733	0.749	0.733	0.727	0.731	0.739	0.724	0.755
Lymphography	0.747	0.727	0.776	0.755	0.741	0.601	0.771	0.688	0.776
M-of-N	0.868	0.836	0.914	0.806	0.837	0.818	0.855	0.921	0.972
Tic-tac-toe	0.732	0.713	0.724	0.673	0.673	0.673	0.737	0.749	0.736
Vote	0.916	0.93	0.912	0.93	0.952	0.966	0.94	0.924	0.932
Zoo	0.855	0.842	0.873	0.655	0.727	0.6	0.885	0.848	0.855

Table 8.5: Experiment results of mean attribute reduction

Dataset	KNN			MI			MI and KNN		
	GA	PSO	GWO	GA	PSO	GWO	GA	PSO	GWO
Breastcancer	0.644	0.644	0.511	0.111	0.178	0.111	0.578	0.667	0.622
Exactly	0.615	0.6	0.492	0.892	0.892	0.892	0.785	0.508	0.508
Exactly2	0.292	0.523	0.235	0.954	0.892	0.969	0.354	0.523	0.262
Lymphography	0.489	0.378	0.127	0.267	0.233	0.0556	0.489	0.489	0.367
M-of-N	0.631	0.569	0.462	0.323	0.369	0.292	0.692	0.585	0.462
Tic-tac-toe	0.6	0.6	0.489	0.111	0.222	0.111	0.556	0.622	0.533
Vote	0.375	0.525	0.413	0.175	0.2125	0.063	0.463	0.363	0.35
Zoo	0.588	0.5625	0.5625	0.2625	0.225	0.063	0.588	0.563	0.45

we can see that GWO has comparable or minimum variance value, which proves the capability of convergence to global optima regardless of the initial solutions, which proves the stability of the algorithm. Also, on the level of best and worst solutions obtained at the different runs we can see advance on the fitness value obtained by GWO over PSO and GA over almost all the test datasets.

Table 8.4 summarizes the average testing performance of the different optimizers over the different datasets. One can see that the performance of GWO is better than GA and PSO for all the used fitness functions over the test datasets. By comparing the performance of different fitness functions used, namely, mutual information, classification performance, and the multi-objective fitness functions, we can see the advance of the proposed multi-objective function on performance. This advance can be interpreted by the good description of data with minimal redundancy and classifier guidance by the second objective of the fitness function.

Table 8.5 describes the average selected feature size by the different optimizers using different fitness functions over the different datasets. We can see that the proposed multi-objective function outputs solutions with minor feature size in comparison to the other single objective fitness functions. Also, we can see that GWO is still performing better for feature reduction.

8.3 Chapter conclusion

This chapter reviewed and discussed the wolf search optimizers and its variants as well as reviewed some real life applications. In addition, it showed some results in how wolf-based search algorithms applied in feature selection problems.

Bibliography

[1] Rui Tang, S. Fong, Xin-She Yang, and S. Deb, "Wolf Search Algorithm with Ephemeral Memory", in: Seventh International Conference on Digital Information Management (ICDIM), 22-24 Aug. 2012, pp. 165-172, Macau, DOI 10.1109/ICDIM.2012.6360147, 2012.

[2] Hu-Sheng Wu and Feng-Ming Zhang, "Wolf Pack Algorithm for Unconstrained Global Optimization", *Mathematical Problems in Engineering*, Vol. 2014, Hindawi Publishing Corporation, Article ID 465082, 17 pages, http://dx.doi.org/10.1155/2014/465082.

[3] Seyedali Mirjalili, Seyed Mohammad Mirjalili, and Andrew Lewis, "Grey wolf optimizer", *Advances in Engineering Software*, Vol. 69, pp. 46-61, 2014.

[4] E. Emary, Hossam M. Zawbaa, Crina Grosan, and Abul Ella Hassenian, "Feature Subset Selection Approach by Gray-Wolf Optimization", in: Afro-European Conference for Industrial Advancement Advances in Intelligent Systems and Computing, Vol. 334, pp. 1-13, 2015.

[5] Ali Madadi and Mahmood Mohseni Motlagh, "Optimal control of DC motor using grey wolf optimizer algorithm", *Technical Journal of Engineering and Applied Sciences*, Vol. 4(4), pp. 373-379, 2014.

[6] W. Yamany, E. Emary, and A.E. Hassanien, "Wolf search algorithm for attribute reduction in classification", Computational Intelligence and Data Mining (CIDM), 2014 IEEE Symposium on Computational Intelligence and Data Mining (CIDM), pp. 351-358, 2014.

9 Bird's-Eye View

In previous chapters, we covered sets of modern numerical swarm-based optimization methods including:

- Bat algorithm (BA)

- Firefly algorithm (FFA)

- Cuckoo search (CS)

- Flower pollination algorithm (FPA)

- Artificial bee colony (ABC)

- Artificial fish swarm algorithm (AFSA)

- Wolf search algorithm (WSA)

- Gray wolf optimization (GWO)

In this chapter, we are trying to classify the above algorithms according to different criteria and find similarities and differences based on these criteria. This discussion may help in finding weak and strong points of each and may help in presenting new hybridizations and modifications to further enhance their performance or at least help us choose among this set to handle specific optimization tasks.

9.1 Criteria (1) Classification according to swarm guide

All swarm intelligence algorithms usually share information among the multiple agents. Thus, at each iteration of the optimization all/some agents update/change their position based on position information of other/own position. If we divided the algorithms based on the number of agents according to which a given agent will change its position, we can divide them into the following:

- *Algorithms following best solution:* BA uses the best solution for updating the velocity vector and hence the agents' positions. ABC uses the global best to update the positions of the experienced bees. WSA uses the best solution inside its visual domain but not the global best.

- *Algorithms following the three best solutions*: GWO uses equally the best three solutions: alpha, beta, and delta, as the target for updating agents' positions.

- *Algorithms following a subset of better solutions*: In ABC, onlooker bees follow one of the experienced bees in roulette wheel manner to update their positions.

- *Algorithms following all better solutions*: FFA is an example of this class where fireflies are moved toward all fireflies that are brighter.

- *Algorithms following random solution(s)*: FPA stochastically changes the position of a given agent as a function of two random agents: local pollination. CS uses two random solutions as a guide for generating solutions; nests, as a replacement of worst solutions.

- *Algorithms following the agents centroid*: Artificial fish swarm algorithm (AFSA) uses the swarm centroid as a guide in the swarm-behavior state.

Figure 9.1 outlines the different swarm intelligence methods division based on the number of agents used in the solution updating.

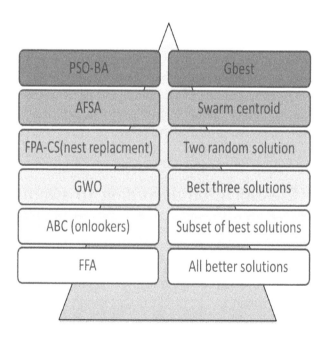

Figure 9.1: Classification of swarm intelligence methods based on the number of agents to follow

9.2 Criteria (2) Classification according to the probability distribution used

On analyzing bio-inspired algorithms, we can see the types of randomness that a particular algorithm is employing. Algorithm uses different randomness generators inside different components of an algorithm, and various probability distributions such as uniform, Gaussian, and Lèvy distributions; Cauchy can be used for randomization. Randomization is an efficient component for global search algorithms and hence a very precious way to add modifications to the basic algorithms. Table 9.1 outlines the different optimization algorithms and the corresponding statistical distribution used inside the algorithm.

We can see that either uniform or normal distributions are commonly used in local searching while Lèvy distribution is more commonly used in global searching.

9.3 Criteria (3) Classification according to the number of behaviors used

Some optimizers use a single formula or behavior to update all agents' positions such as PSO, where a single equation is used to update the agent's velocity and hence the agent's position. FFA uses a single formula to update the agent's position where this formula incorporates attraction by other better fireflies and environment modeling.

In BA two behaviors are used to update bat's position, namely, local move and global move. In the local move the bat searches its surrounding to find a better solution using normal random search around the global best. In the global searching behavior bats update their position according to the position of the current global best solution.

FPA uses also two behaviors for searching, namely, local pollination and global pollination. In the local pollination FPA two random agents are used in the pollination. In the case of global pollination, FPA used Lèvy flight to move a pollen toward the global best solution.

CS uses two different formulas to update its agents' positions: global random walk and local random walk. The local search is performed around the global best using Lèvy distribution. In the global search, a fraction of agents are updated by selecting two random agents and using them to update the agent's position.

ABC uses three different formulas for position updating and each is dedicated to one class of bees: scout, experienced, and onlooker bees. The experiencd bees update their position using the guidance of the global best. An onlooker bee selects one of the experienced bees as a guide for position updating. In the scout bee class, the agent randomly updates its position.

Table 9.1: Optimization algorithms and the used statistical distribution

Optimizer	Uniform Distribution	Normal Distribution	Lèvy Distribution
PSO BA	Velocity updating • Switching between local and global search • Stochastically moving an agent to a better solution	In local search	
CS	• Local random walk • Selecting agents for local search • For finding nests to be abandoned		The global random walk
FFA	Modeling the environment randomly	Modeling the environment randomly	
FPA	• Switching between local and global pollination • Local pollination		Global pollination step modeling
ABC	• Updating the position for experienced bee which balances positions change by bee's own experience and swarm experience • Controls the attraction of the onlooker bee toward its interesting food source area	Local search in the scout bee updating	
AFSA	• In guiding the search in the random, searching, swarming, and leaping behaviors of fish • Selecting random agent in the searching behavior and leaping behavior of fish		
WSA	• Position updating in Escape behavior and passively preying behavior and initiatively preying behavior • Deciding escaping wolves		
GWO	Position updating using alpha, beta, and delta solutions		

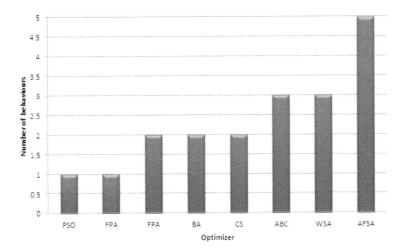

Figure 9.2: Classification of SI methods based on the number of behaviors used in agent's repositioning

WSA switches among preying initiatively, preying passively, and escape behavior. In the preying initiatively, the wolf is attracted by the best wolf position in its visual domain. Preying passively uses random local search around individual agents to reposition the wolves. In the escape behavior the wolf goes outside its visual domain randomly.

AFSA uses either random, search, swarm, chase, or leap behavior to update agent's position. The switching among these states is controlled by the crowd parameter to avoid algorithm stagnation. In the random behavior, an agent updates its position using a random position. In the searching behavior, a fish updates its position with the guide of a random agent inside its own visual range. In the swarming behavior, a fish is moved toward the swarm centroid. The chasing behavior moves a given fish toward the current global best solution. The leaping behavior is applied by repositioning a given agent in the search space randomly. Figure 9.2 outlines the number of behaviors used inside the different optimizers.

By increasing the number of behaviors in the optimization algorithm we increase the complexity and running time of the optimizer but, on the other hand, it enhances searching the solution space and helps avoid premature convergence.

9.4 Criteria (4) Classification according to exploitation of positional distribution of agents

AFSA is the optimizer with most exploitation of agents' positions in deciding the behavior and hence the position updating mechanism. Crowd parameter which estimates the number of search agents in the visual scope of a given agent is used to decide the

FPA, WPA, GWO, CS, ABC and
BA

Figure 9.3: Hierarchical representation for exploiting spatial distribution of agents in the swarm algorithm

fish behavior. If the crowd parameter is high the fish applies search behavior, while if it is empty the fish switches to random behavior; otherwise the fish uses swarm or chase behavior.

In the wolf search algorithm uses similar behavior as AFSA where it uses the visual scope in the optimization. In the preying initiatively behavior, the agent follows the best agent in its visual scope and goes to preying passively behavior if none of the wolves in its visual scope is better.

FFA implicity used the spatial distribution of other fireflies where its distance between firefly i and firefly j is a major component in updating the ith firefly position.

Other algorithms such as FPA, WPA, GWO, CS, ABC, and BA do not monitor the distribution of other agents at all. Figure 9.3 outlines the hierarchy of exploiting spatial agents' distribution into the swarm algorithm.

Naturally, the exploitation of spatial information helps the algorithm to avoid stagnation and hence helps better exploration of the search space, but it may require much more time to converge.

9.5 Criteria (5) Number of control parameters

All meta-heuristic algorithms have algorithm-dependent parameters, and the appropriate setting of these parameter values largely affect the performance of the optimization. One of the very challenging issues is deciding what values of parameters to use in an algorithm at a given problem. How can these parameters be tuned so that they can maximize the performance of the algorithm of interest [1]? Parameter tuning itself is an optimization problem.

There are two main solutions for adapting a given parameter set for an optimizer, namely, trial and error and by experience. In the trial-and-error approach, the algorithm is run and the parameters are adapted until reaching a satisfactory result. Some experts in a given operational field and given an optimization algorithm can set the control parameters using their own experience and exploiting their knowledge about the search space and shape of the objective function. A third approach was presented by Yang in [1] using an algorithm (which may be well tested and well established) to tune the parameters of another relatively new algorithm.

By analyzing the optimizers discussed here, we can see that the GWO is the optimizer with almost 0 parameter to be tuned; $\vec{a}$ controls the trade-off between exploration and exploitation. The creators of that algorithm proposed an initialization and adaptation for this parameter but this doesn't prevent other methods from adapting and initializing it.

In CS, the discovery rate parameter is a critical parameter as it controls the rate of abandoning a given nest. Higher values for this parameter allow for much more exploration but of course slow down the convergence speed of the optimizer. On the other hand, lower values for this parameter allow for much exploitation but the threat of falling in a local minimum is high. Thus, a good choice should be selecting higher values for this parameter at the begining of optimization and at later iteration this parameter should be decremented. Another effective parameter in CS is the β parameter that controls the shape of the Lèvy distribution used. This parameter is always set to constant value.

In BA, six parameters are the key ones, namely, loudness, pulse rate, pulse rate adaptation (γ), loudness decrement (α), local search radius, and frequency band parameters. Frequency band parameter controls the amount in change in bat's velocity and hence low band limit can reduce the convergence speed of the algorithm but ensures intensive searching of the search space. Higher values for this upper limit of this parameter allow for faster convergence but with less intensification. The local search parameter should start with large values at the begining of optimization to allow for much exploration and should be decremented at later iterations to allow for intensive locals searching. The loudness parameter decides the rate of catching a prey, where if loudness is high and better solution is caught by a given the bat can neglect it and keeps its own position. On the other hand, at lower loudness values a given bat always catches any better solution. Of course this parameter is set to high values at the begining of optimization and decremented by a rate α. The last effective parameter in BA is the pulse rate parameter which controls the rate of local searching. Of course local searching rate increases as optimization advances and hence this parameter should start with small values and increases with time at a rate γ.

In FFA, the parameter α or randomness weight parameter is used to model noise in the environment around the firefly swarm. This parameter affects the performance of FFA as it adds a random component to the firefly step. For fitness functions with many local minima this parameter should be set to high values while in fitness functions with smooth terrain it should be set to low values. Higher values for this parameter slows down convergence but increases the chance for extra exploration. On the other hand, low values for this parameter allows for much speed convergence but much change for swarm stagnation. This parameter should start at high values and decrements as optimization proceeds given a rate δ.

The attraction strength is controlled by the parameter β_0 that controls the amount of attraction by other brighter fireflies. At higher values for this parameter the given firefly can be named an onlooker firefly that only follows other fireflies, while in the case of lower values for this parameter the firefly is said to be very selfish, that is, it uses its own experience. At lower values for this parameter the firefly is very dedicated to local searching and hence can quickly converge but at possible local minima. On the other hand, at high values for this parameter the firefly has more explorative capability but it slows down convergence.

In FPA, the parameter p that controls the switch between local and global pollination is an effective one. At high value for this parameter the algorithm is much tending toward exploration and hence requires much time to convergence. At lower values for this parameter the FPA applies many steps in exploitation and hence it is liable to premature convergence. This parameter should be set to high values at the begining of optimization and hence decremented to reach its lower bound at the end of the optimization. Another effective parameter in CS is the β parameter that controls the shape of the Lèvy distribution used. This parameter is always set to constant value.

In ABC, the parameter that controls the ratio of scout, onlookers and experienced bee is very effective. Swarms with many scout bees are very explorative and require longer time to converge. At the other end swarms with much experienced bees are converges faster but at probably local optima. Thus, the number of experienced bees should be lower at the begining of optimization to allow for many scout explorative bees, while at the end of optimization the experienced bees should allow more for exploitation and intensive searching. Naturally, a parameter that controls the trade-off between its own experience and better bee experience is another effective one. In the case of using its own experience more the bee swarm becomes more cautious in its steps and hence it becomes more exploitative. On the other hand, on using lower values for this parameter the bees are much attracted to the better location and hence allow for faster convergence.

In AFSA, the visual domain is the most important parameter as it controls the behavior of a given fish. Fish with low visual domain become more exploitative while fish with larger visual domain are more explorative. The crowd parameter is also very effective in performance. Lower values for crowd parameters allow for much application of the searching behavior that allows for exploration. On the other extreme, at higher values for crowd parameter the chasing and swarming behaviors are more applicable and hence allow for much faster convergence. Thus, this parameter should be set to low values at the begining of optimization and increases as optimization proceeds.

In WSA, the parameter Pa that controls the threat probability is an effective parameter that is used to stochastically apply the *escape* behavior. Too much application of *escape* behavior allows for much exploration but slows down the convergence, while, on the other hand, when the *escape* rate is very small the algorithm devotes much effort in exploitation and hence speeds up convergence. So this parameter should be set to high value; to allow for exploration, at the begining of optimization and then decreasing in time. The α parameter controls the radius of local searching. This parameter should be set to higher values at the begining of optimization and decreases in time.

9.6 Criteria (6) Classification according to either generation of completely new agents per iteration

Weak agents are replaced with better or random ones in some algorithms, while other algorithms always keep their agents forever. In CS a ration of nest is discovered and destroyed at each iteration and new nests are created randomly. In WPA, a fraction of wolves with worst fitness are replaced by new random ones. While in AFSA leaping behavior is applied when no advance is achieved in the fitness, while when no fish exist in the fish's surround, visual scope, the fish applies random behavior.

Scout bees in AFSA perform the task of renewing the swarm where scout bees always perform random search to explore new search areas. Other algorithms such as GWO, BA, PSO, and FPA don't explicitly change the swarm but depend on updating the positions of each agent individually. Performing the renewing of swarm always leads to much exploration power and is used as a solution to the well-known stagnation problem. Figure 9.4 is a hierarchical representation of algorithms and their swarm-renewing strategies.

9.7 Criteria (7) Classification based on exploitation of velocity concept in the optimization

Velocity is change in position through time. Velocity concept is explicitly used in optimization in some methods such as BA, while it is implicitly used in other algorithms.

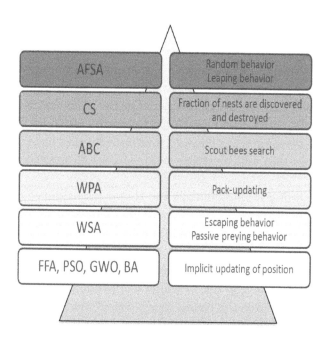

Figure 9.4: Hierarchical representation for different strategies for renewing the swarm in optimization

Velocity updating in PSO exploits the particle's best position and the global best position and hence can be considered as a generalization for the velocity updating in BA, which exploits only the global best position. Other algorithms implicitly use velocity for position updating. In FFA velocity is represented by two components: one is related to attraction and the other is related to environment modeling and randomness. In FPA velocity in the global pollination is much variant as it makes use of Lèvy flight, which is very systematic as in local pollination. WSA uses different velocities according to its current behavior where it can prey initiatively, prey passively, or escape. In WPA velocity is controlled by current behavior such as scouting, calling, besieging behaviors. CS uses Lèvy flight for updating the velocity implicitly and hence provides for much variation in the velocity. In ABC the experienced bee uses its own best and global best to update velocity, while onlooker bees use selected elite bees to update their position. AFSA changes position according to the current behavior and uses systematic updating except for leaping behavior that has abrupt change in position.

Figure 9.5 describes the velocity updating where velocity may be defined explicitly as in BA. Implicit definition of velocity may use a direct deterministic equation for calculating the amount of change in position as in FFA and WSA. Stochastic walks are used also based either on random walks as in FPA, WSA, GWO or based on Lèvy flight as in FPA and CS or based on Browning motion as in BA and ABC.

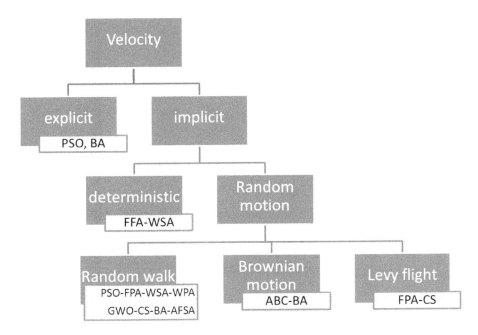

Figure 9.5: Hierarchical representation for different strategies for renewing the swarm in optimization

9.8 Criteria (8) Classification according to the type of exploration/exploitation used

Exploration can be defined as acquisition of new information throughout searching [2]. Exploration is a main concern for all optimizers as it allows for finding new search regions that may contain better solutions. Exploitation is defined as application of known information. The good sites are exploited via the application of a local search. The selection process should be balanced between random selection and greedy selection to bias the search toward fitter candidate solutions (exploitation), whilst promoting useful diversity into the population (exploration) [2]. The searching may be along a given direction (diversification), around a specific point (intensification), or completely random (diversification) in the space. Combinations of the past three strategies for search are common in the SI algorithms.

- *Searching along a direction:* The updating of a given agent position takes the generic form

$$x_i^{t+1} = x_i^t + A\varepsilon(x_1^t - x_2^t) \tag{9.1}$$

where x_i^t is the position to be updated, x_1^t, x_2^t are two solutions from the swarm, ε is a constant always random controlling the position along the selected direction, and A is an adaptable constant updating speed, that may depend on the problem and/or the optimization iteration number.

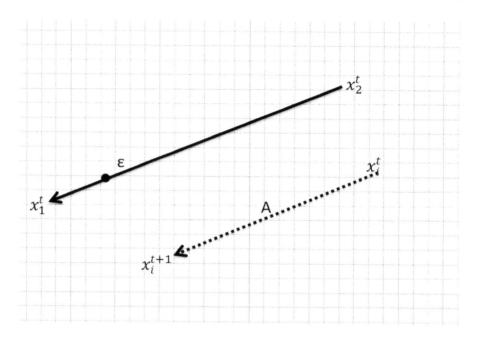

Figure 9.6: Updating of agent x_i^t position in the direction $x_2 x_1$

Figure 9.6 is a graph representing the effect of the updating direction $\overrightarrow{x_2 x_1}$ on the position of agent x_i^t.

All optimization algorithms in the study make use of this type of searching. For example, in PSO the A parameter is called loyalty factor and is always set according to the problem in the range from 0 to 1, and ε is selected as a uniform random number $\in [0, 1]$. x_1, x_2 are particle's best solution and the agent's position in order. The same notation is incorporated in the PSO updating but x_1 replaced by the best solution is also used.

In FFA the A is named β_0 and is always set to 1, ε is represented as $e^{-\gamma r_{ij}^2}$ where r_{ij} is the distance between fireflies i and j, while x_1, x_2 are the brighter and darker fireflies in order.

FPA uses Lèvy flight as a replacement for ε and $A = 1$, with x_1, x_2 the flower's pollen and the best pollen in global pollination. In case of local pollination ε is a random number drawn from uniform distribution in the range [0,1], $A = 1$, and x_1, x_2 are two random pollens.

In WSA, the A factor is set to 1 and ε is set to e^{-r^2} with r being the distance between $wolf_i$ and the best wolf in its visual domain. x_1, x_2 are set as x_i^t, x_{best}^t and

x_{best}^t is the wolf with best fitness in the visual domain of the wolf positioned at x_i^t at iteration t.

WPA uses A as 1 and ε is the random step size, while x_1, x_2 are the positions of the lead wolf and the wolf's positions x_i^t in the calling behavior.

In CS, the main updating equation uses $A = 2$ and ε as a random number drawn from Lèvy distribution. x_1, x_2 are used as agent's old position and best agent position in order. When abandoning a nest CS used another form of updating: x_1, x_2 are two random nests' positions and ε a random uniform number in the range [0,1] and A is 1.

In foraging bees updating in the ABC, the A parameter is called loyalty factor and is always set according to the problem in the range from 0 to 1, and ε is selected as a uniform random number $\in [0, 1]$. x_1, x_2 are the bee's best solution and the agent's position to be updated in order. Also same notations are used with x_1 as the global best bee position. In onlooker bees, the A parameter is called loyalty factor and is set $\in [0, 1]$, and ε is selected as a uniform random number $\in [0, 1]$. x_1, x_2 are elite or experienced bee's position and the agent's position to be updated in order.

In BA, the bat's velocity is updated using the same equation with $A = 1$, ε set as frequency, and x_1, x_2 are the bat's position to be updated and position of the best bat in order.

In AFSA, the searching along a direction is changed according to the fish's current behavior, but all use $A = 1$, ε as uniform random number $\in [0, 1]$ and x_2^t is set as current fish's position to be updated. x_1^t is set as the position of a random fish, the visual domain of x_2^t in the case of searching behavior, as swarm centroid in case of swarming behavior and as the global best fish position in the case of chasing behavior.

- *Searching around specific solutions:* This search is always performed for exploitation and has the following generic form:

$$x_i^{t+1} = x_i^t + \alpha rand \tag{9.2}$$

where x_i^t is the solution to be updated, α represents scaling factor, and *rand* is a random number. In FFA fireflies use this as a random or mutation component that models environment purity, α is a constant in the range from 0 to 1, and rand is set as a uniform random number $\in [-0.5, 0.5]$.

Table 9.2: SI algorithms and the corresponding search direction

Method	*Search Direction*
Generic form	$x_i^{t+1} = x_i^t + A\varepsilon(x_1^t - x_2^t)$
PSO	$gbest - x_i^t$ and $pbest - x_i^t$
FFA	$x_j^t - x_i^t$
FPA	$x_{best} - x_i^t$ or $x_j^t - x_k^t$ two random pollens
WSA	$x_{best} - x_i^t$ with x_{best} is the best wolf position in the visible domain of x_i^t
WPA	$x_{best} - x_i^t$ with x_{best} is the leading wolf position
CS	$x_{best} - x_i^t$ or $x_j^t - x_k^t$ are two random nests
ABC	$gbest^t - x_i^t$ and $pbest - x_i^t$ for experienced bees or $x_elite^t - x_i^t$ with x_elite one of the experienced bee for onlooker bees
BA	$gbest^t - x_i^t$ with $gbest$ is best bat position
AFSA	$x_c^t - x_i^t$ in swarming behavior or $x_{best}^t - x_i^t$ in chasing behavior or $x_{rand}^t - x_i^t$ in search behavior or

In WSA α is set to a constant $\in [0, 1]$ and *rand* is a uniform random number $\in [0, 1]$ and this factor can be considered as a mutation term to help escape from local minima.

In the WPA α are set to 1, while rand is set as $sin(2\pi * \frac{p}{h}) * step_a^d$, where p is the dimension number and h is a random integer between 1 and p.

In ABC scout bees use this type to explore new regions, with *rand* named *RW* as a random walk function that depends on the current position of the scout bee, and α named the radius search ϑ.

BA uses the same formula stochastically and performs searching with α as average loudness and *rand*, called ϵ, is a uniform random number $\in [-1, 1]$.

In AFSA the leaping behavior is performed using this type of search, with *rand* in the position of a random fish in the swarm and α is set to a constant > 0.

- *Complete random position:* This type of search is used to perform abrupt jumps in the search space and has the following form:

$$x_i^{t+1} = x_i^t + C \tag{9.3}$$

where x_i^t is the agent's position to be updated and c is a constant. The only optimizer using this form is the WSA in the escaping behavior where C is set as the visual domain range to allow the wolf to escape from threats.

- *Other:* In GWO another updating form is applied not using any of the above three main types.

9.9 Chapter conclusion

In this chapter, we gave a bird's eye view of the modern swarm optimization techniques. In addition, we provided several criteria that are used to differentiate among these swarm techniques. It identifies similarities and differences in the criteria setting. This view may help in finding weak and strong points of each algorithm and may help present new hybridizations and modifications to further enhance their performance or at least help us choose among this set to handle specific optimization tasks. We believe that the future in swarm optimization will be as active as the past and will bring many advances in swarm optimization and chagrin with many approaches currently used in practices.

Bibliography

[1] Xin-She Yang, Zhihua Cui, Renbin Xiao, Amir Hossein Gandomi, Mehmet Karamanoglu, *Swarm Intelligence and Bio-Inspired Computation; Theory and Applications*, Elsevier, Waltham, Mass. 2013.

[2] Xin-She Yang, Nature-Inspired, Metaheuristic Algorithms, 2010, Luniver Press, Bristol, UK.

Index

Swarm Intelligence: Principles, Advances, and Applications